Statistical Process Control
A Pragmatic Approach

Continuous Improvement Series

Series Editors:

Elizabeth A. Cudney and Tina Kanti Agustiady

PUBLISHED TITLES

Affordability: Integrating Value, Customer, and Cost for Continuous Improvement
Paul Walter Odomirok, Sr.

Continuous Improvement, Probability, and Statistics: Using Creative Hands-On Techniques
William Hooper

Design for Six Sigma: A Practical Approach through Innovation
Elizabeth A. Cudney and Tina Kanti Agustiady

FORTHCOMING TITLES

Transforming Organizations: One Process at a Time
Kathryn A. LeRoy

Statistical Process Control: A Pragmatic Approach
Stephen Mundwiller

Robust Quality: Powerful Integration of Data Science and Process Engineering
Rajesh Jugulum

Building a Sustainable Lean Culture: An Implementation Guide
Tina Agustiady and Elizabeth A. Cudney

Statistical Process Control
A Pragmatic Approach

By
Stephen Mundwiller

CRC Press
Taylor & Francis Group
Boca Raton London New York

CRC Press is an imprint of the
Taylor & Francis Group, an **informa** business

CRC Press
Taylor & Francis Group
6000 Broken Sound Parkway NW, Suite 300
Boca Raton, FL 33487-2742

First issued in paperback 2020

ISBN-13: 978-1-4987-9913-3 (hbk)
ISBN-13: 978-0-367-78177-4 (pbk)

Library of Congress Cataloging-in-Publication Data

Names: Mundwiller, Stephen, author.
Title: Statistical process control : a pragmatic approach / Stephen Mundwiller.
Description: Boca Raton : CRC Press, Taylor & Francis Group, 2017. | Series: Continuous improvement series | Includes bibliographical references.
Identifiers: LCCN 2017019769 | ISBN 9781498799133 (hardback : acid-free paper)
Subjects: LCSH: Process control--Statistical methods.
Classification: LCC TS156.8 .M844 2017 | DDC 660/.2815--dc23
LC record available at https://lccn.loc.gov/2017019769

Visit the Taylor & Francis Web site at
http://www.taylorandfrancis.com

and the CRC Press Web site at
http://www.crcpress.com

To Dr. John Ridgway, PhD, University of Missouri–St. Louis,
who told me in freshman biology class, "Steve, if you earn
this degree, you will have learned how to think."

To Dr. Lon Wilkens, PhD, University of Missouri–St. Louis, who
told me in 2006, "It is never too late to right a wrong."

To Dr. Elizabeth Cudney, PhD, Missouri University of Science and
Technology, for her faith in me and her support in writing this book.

To my wife Deborah, for her support and encouragement.

To my daughter Stephanie, for just being wonderful.

Contents

Acknowledgments

I am grateful and indebted to those who have taught and mentored me over many years. They may have had just a few kind words or definite examples and outright instruction. They challenged me to think, although sometimes in a painful way, but always to a positive end. Sometimes, the thought process from their challenge occurred years later, but nevertheless it did happen.

About the Author

Stephen Mundwiller is currently employed by Liebel-Flarsheim, LLC, a Guerbet Group Company. Prior to this, Stephen performed consulting and training services through his own company, SME Quality Resources, LLC.

In 2014, Stephen left his position as the director of quality assurance and regulatory affairs for Allied Healthcare Products in St. Louis, Missouri. In this capacity, Stephen was responsible for improving and maintaining the quality management systems, ensuring the timely submission of all domestic and international regulatory documents, managing the investigation of customer complaints, managing quality improvement activities, and managing a staff of 13.

Previously, also through SME Quality Resources, Stephen provided comprehensive consulting services, training and mentoring services, and systems development to manufacturing and service sectors, both public and private. He has served many organizations by implementing and improving quality systems, mentoring and providing instruction to individuals, and by performing quality system audits.

Currently, Stephen serves the American Society for Quality (ASQ) as the Deputy Region 13 Director. He served as a board member of the St. Louis chapter of the ASQ for approximately 10 years and holds ASQ certifications as a Manager of Quality and Organizational Excellence (CMQ/OE), Quality Systems Auditor (CQA), Quality Process Analyst (CQPA), and Six Sigma Green Belt (CSSGB). In addition, Stephen has been an instructor for St. Louis Community College, teaching classes in ISO 9000 and Quality Tools.

1

Why Statistical Process Control?

Control without action is simply a hobby.

Kaoru Ishikawa

Why? Because it will still work and it works great when used properly, with appropriate training. Just because something is old, does not mean it no longer provides value. In this Six Sigma world, there are other methods to monitor, control, and improve processes. Since Six Sigma is basically a project management methodology, it can be used to reduce or eliminate issues identified by statistical process control (SPC) charts. Note that as you read this book, I am a pragmatic fanatic when it comes to product or service quality.

SPC does not solve anything. SPC charts provide data, and extremely meaningful data if one understands what they are seeing when they look at the control chart. That data provides a picture of products that have been manufactured or a service that has been provided. At this point, I must state that I am a manufacturing guy. While I've consulted in the service industry, approximately 98% of my background is in manufacturing. So this book, the examples, and the war stories provided are from manufacturing scenarios.

Years back, I worked as a quality manager in a high-speed liquid, consumer product manufacturing factory. Line speeds were generally 80–150 bottles per minute depending on the size and the viscosity of the liquid. The lines were mostly automated with a crew of three to five people per line. It was a three-shift operation with occasional weekend work, although not all lines ran on second or third shift. The hourly workers were all trained in SPC, mostly by me. The first thing I did when I arrived at my desk in the morning was to turn on my PC and review the SPC data from the previous second and third shifts. With the SPC software used and essentially any SPC software on the market, one can isolate the data to a production line, a shift, or any time period. Based on the charts, I could predict their shift's level of quality, their approximate case output, and what mistakes, if any, were made, as well as the level of operator frustration. But, then, I'm highly talented. Although with time and training almost any professional could do the same. I'm just pointing out the value of accurate SPC data with accurate control charts.

During this time, I could identify remedial training needs by evaluating the SPC control charts. An example would be when the primary line leader was on vacation and the backup employee was running the line. Sometimes, it was a case of nerves. Other times, they had clearly forgotten some of their SPC training. While not a catastrophe, I could easily address these special or assignable causes of variation. Later in this book, I will discuss what I call problem variation, which I consider the most important type of variation to be able to recognize.

My SPC training classes started in this organization with the first-shift production lines of which there were about a dozen, as all lines usually ran on first shift. After first shift was trained, I then worked second shift to provide training to the smaller second shift, which only ran a few lines based on demand. Their training was followed by the third shift. This training was a two-year ordeal, as training was not the primary focus of my job. Like many production scenarios, taking hourly direct labor personnel out of producing product is almost unheard of and not something done routinely.

With the first shift trained, there was a noticeable reduction in variation as shown on the SPC control charts. Two main reasons: data entry errors were able to be edited, and I constantly preached to avoid machine or line adjustment unless it was absolutely necessary. If the process was in control and the bottle fill levels, as well as the cap off-torque, met or were close to the minimum, then my policy was to let the line run. More on this later, but as stated, adjusting a process is adding a variance, and things will get worse before they get better.

So when untrained second shift operators, who tended to be younger and male, took over the production line from the more experienced and SPC-trained first-shift operators, who tended to be female, their first thought was, "I can get out more cases on my shift than she did!" So what did they do? They increased the line speed, which added a variation and was an assignable cause. The filler became out of synchronization, the capper and labeler started jamming, and the case packer would malfunction. So they would have to stop the line many times. Their case output would be about 60% of the first shift. Why? They added variance to the process by making adjustments. Not really their fault, as they did not know any better until I had trained them. I will repeat this multiple times in this book. When making changes to a process, things will always get worse before they improve, albeit sometimes for a short period of time. Once I had second and third shifts trained, these types of competitive issues went away. On some lines, engineering and I had the adjustments locked into a specific setting where no adjustments could be made except by maintenance. Overall, case output went up significantly for the entire operation.

I had other very significant accomplishments at this facility. Eventually, they got into very serious financial trouble. They had to cut staff and as in many instances, quality assurance (QA) is the first to go. Such is life for the hard-working quality manager.

Other special or assignable causes that are seen on the control charts would fall into one of the six categories as listed on a Cause and Effect Diagram or Fishbone Diagram.* These are environment, method, materials, measurement, manpower, and machine. This would provide projects for quality, maintenance, engineering, or sometimes human resources. Without the control charts, it would have been much more difficult to identify these special causes of variation unless there was clear catastrophic failure. One thing that should be noted is that in high-speed manufacturing like this, stopping the production line is detrimental to the shift case output and adds unnecessary variation to the process. Yes, stopping a production line is adding variance to the process.

At other times, we used the SPC control charts to identify more long-term projects to reduce the common cause variation inherent in the process. While I will discuss this more in Chapter 4 on variation, think of it as the background noise of the process. Life is not perfect, manufacturing is not perfect, service is not perfect, and control charts are not perfect in appearance. This is shown on the control chart as the common cause variation.

Another advantage of having an SPC program in place is that it enables one to quantitatively measure the reduction in variation after an improvement is implemented. Again, always keep in mind that when a change or changes are made to a process, that is introducing a new variable. Things will get worse before they get better. This will be shown on the control charts. So

---

* Cause and Effect Diagram, Ishikawa Diagram, or Fishbone Diagram: Developed by Kaoru Ishikawa (1915–1989) in 1968, consisting of six segments or branches on the diagram: machine (technology), method (process), material (raw material and consumables), manpower (people), measurement, and environment. When drawn, the diagram resembles a fishbone.

don't panic. Be patient! Again, later in this book I will discuss what I call problem variation.

For example, a new machine is added to a production line to replace an old one that needs excessive maintenance. The new machine is also capable of higher capacity. When the production line is restarted after that new machine has been installed, production will be slower and quality will likely be worse. The new machine has to be "fine-tuned." It has to be adjusted to be compatible with other pieces of equipment on the production line. Other machines on the production line may need adjustment to be more compatible with the new machine. Then let the production line run. Let things start to synchronize. Don't panic and start over adjusting and blaming the new machine. Take it slow and easy. Gradually, the process will get better. The control charts will graphically show this slow improvement as a slow reduction in common cause variation. The variation caused by putting in the new machine is special or assignable cause variation. If the common cause variation levels out with no real overall improvement, then the new machine was not the answer to reduce the common cause variation. However, there is still an advantage due to the reduction in maintenance and downtime. All this is seen by using SPC and viewing the control chart. Note again that every time a production line stops and then starts there is an introduction of special cause variation which could be a source of problems.

While an engineer or quality professional may accumulate data and construct a control chart manually, the best means to track the SPC data is with software. Whether the measures are taken manually or automatically, those measures are entered onto an SPC control chart or into an SPC system. Manual SPC charts may be used to evaluate a specific set of data over a brief period of time if SPC software is not available. Clearly, software is the best option for a production scenario. Every time new data is entered, the software does the calculations … we'll discuss these calculations in Chapter 3. The software will hold tens of thousands of pieces of data. The data present can cover years of production. Or one can isolate and just look at any specific time period. This is very useful to evaluate the differences between shifts and operators, identify improvement projects, or measure the value of improvement projects and the level of improvement.

For all practical purposes, SPC does not work in a real world sense unless SPC software is used. It should be noted that the software will immediately perform all of the various calculations and update the important numerical indices, as well as the control charts. This provides the user, management, and any other interested parties with an immediate picture of the process performance.

Today, we have Six Sigma, Lean, and Lean-Six Sigma as the solution to all things in the quality and business world. I consider Lean or Lean Manufacturing to be a cultural change and Six Sigma to be a project management methodology. As I've discussed in this chapter and will in more detail in the next, SPC is not new and it is actually rather old in the time frame of the modern manufacturing world. But, it still works and will work extremely well. It can be used in conjunction with modern tools such as Lean and Six Sigma to significantly provide business improvement. While improvements can be made without SPC as I've done as a consultant many times, it is quite useful when used together with Lean and Six Sigma or other methodologies. As shown in Chapter 4, SPC can be used to identify projects or processes needing improvement and then when completed, assess the level of improvement that was accomplished. It should be noted that SPC is a monitoring tool. It is something that provides warnings or clear indications when there is a problem that needs to be addressed.

When an organization goes through a cultural change to Lean, it is truly a change in the way an organization functions. There are eight classic wastes identified that need to be eliminated or greatly reduced as an organization goes through this cultural change. Reducing or eliminating waste releases capacity. One of these classic wastes to work on is to eliminate or reduce defects. Consider how many problems defects cause. Scrap, overtime, rework, excessive standard labor hours, excessive engineering hours, excessive quality control hours, and reduced efficiency are the main consequences of defects. While SPC cannot be applied to all activities involved to reduce the defect rate, it can be used to monitor a process to determine if the common cause variation, assignable cause variation, and what I call problem variation is in fact reduced or eliminated.

Today, we truly live in a Six Sigma world. Every organization wants belts. Green Belts, Black Belts, and Master Black Belts are typical quality and engineering job requirements. SPC can definitely and should be used with Six Sigma projects. Whether by manually calculated results and manually plotted control charts or by using SPC software, SPC is a great tool to measure the progress and success of a Six Sigma project. Not all Six Sigma projects, but certainly some, can be used with SPC for improving a process or business. Unfortunately, not very many Six Sigma professionals use or understand SPC. I've seen multiple holders of Six Sigma black belts that could not interpret a simple control chart. Remember, SPC in itself does not solve problems. SPC is only used to support a process. While that may be for long-term monitoring and just keeping an eye on things, or to identify projects to work on, or to monitor the progress of an improvement, SPC remains a valuable tool, albeit a rather old one.

2

A Brief History

You cannot inspect quality into a product.

Harold F. Dodge

In 1939, while working for Bell Telephone Laboratories, Inc., Dr. Walter A. Shewhart, PhD, wrote *Statistical Method from the Viewpoint of Quality Control*.* The book (currently available from Dover Publications) is based on a series of four lectures Shewhart gave on the subject to the Graduate School of the Department of Agriculture. In the preface, Shewhart noted that statistical methods of research had been highly developed in the field of agriculture. He similarly noted that statistical methods of control had been developed by industry for the purpose of attaining economic control of quality of product in mass production. Shewhart stated that it was reasonable to expect that much is to be gained by correlating as far as possible the development of the two types of statistical techniques.

In his book, Shewhart explained the statistical mathematics behind control and the methodology for plotting data on a control chart. The calculations for the process average, the upper and lower control limits were described and demonstrated. He also included such concepts as the minimum and maximum or the range of the data, subgroup average, standard deviation, and the quantity of data.

Contributors to Dr. Shewhart's book included Dr. W. Edwards Deming, PhD,[†] and Harold F. Dodge.[‡] Dr. Deming always referred to control charts as Shewhart charts in reference to the original developer.

For decades, Dr. Deming preached data-based decision making. Sounds great, doesn't it? But, Deming also stated that using statistical methods alone is not enough. His goal in life was to seek methods of improvement.

---

* Dr. Walter A Shewhart (1891–1967): Wrote *Statistical Method from the Viewpoint of Quality Control*. Originally published in 1939 by the U. S. Department of Agriculture in Washington D.C. Reprinted in 1986 by Dover Publications, Inc.

† W. Edwards Deming, PhD (1900–1993): An American engineer, statistician, professor, author, lecturer, and management consultant.

‡ Harold F. Dodge (1893–1976): An American engineer and scientist and the principal architect of the science of statistical quality control, who is universally known for his work in developing acceptance sampling plans.

Deming's Fourteen Points* for management are still extremely applicable today.

There were 10 or fewer points for good management when he originally wrote the points for management while working in Japan. After returning to the United States, he realized that due to cultural differences in management and the workforce, additional points were needed. For example, point eight, to drive fear out of the workplace, was not needed in Japan due to the culture.

The Fourteen Points

1. Create a constancy of purpose for improvement of product and service.
2. Adopt a new philosophy.
3. Cease dependence on mass inspection.
4. End the practice of awarding business on price alone.
5. Improve constantly and forever the system of production and service.
6. Institute training.
7. Institute leadership.
8. Drive fear out of the workplace.
9. Break down barriers between staff areas.
10. Eliminate slogans, exhortations, and targets for the workforce.
11. Eliminate numerical quotas.
12. Remove barriers to pride of workmanship.
13. Institute a vigorous program of education and retraining.
14. Take action to accomplish the transformation.

A few years later, Dr. Deming had what he termed a "later awakening." He then proposed the "Seven Deadly Diseases"† as he continuously honed his principles for management and business.

---

* *Out of the Crisis* by Dr. Deming: Published by MIT Press in 1982, which included his Fourteen Points for Management.
† Seven Deadly Diseases of Management: Dr. Deming had what he termed as a later awakening after he developed the 14 Points. The first five Deadly Diseases were global in nature while diseases six and seven were for the USA.

The Seven Deadly Diseases

1. Lack of constancy of purpose
2. Emphasis on short-term profits
3. Evaluation by performance, merit rating, or annual review of performance
4. Mobility of management
5. Running a company on visible pictures alone
6. Excessive medical costs (Western countries only)
7. Excessive costs of warranty, fueled by lawyers that work on contingency fees (USA only)

Consider number four, the mobility of management. I once attended a symposium featuring Mr. Charles "Chuck" Knight* of Emerson Electric as the keynote speaker. He was the chief executive officer (CEO) and chairman of Emerson for over 25 years. In his keynote address, Mr. Knight discussed how a CEO or senior leader must stay with a company for at least 20 years in order to accomplish any items of substance. He blamed the short-term tenure of senior managers for the failure of many organizations. Looking at the success of many organizations, consider how many have had long-term leaders.

This is not a book about Dr. Deming or any other guru. My point in listing the brilliant information from Dr. Deming and others is that control charts and statistical process control (SPC) are only a slice of the pie for process improvement and defect elimination. Lean, Six Sigma, formal quality systems, total quality management (see the book by Dr. Dale Besterfield, PhD†), quality circles, and world-class manufacturing techniques, as well as many other quality tools, are some other methods that can be used in conjunction with SPC. SPC provides the picture and the data calculations, but not the solution. For example:

> I call my friend Mike and say, "Hey, how about lunch, are you busy today?" If his response is, "Steve, my line number three is running today with a Cpk of 0.8." My quick response would be, "I'll let you go, you've got stuff to do."
>
> Then, if at a later date I make the same invitation and his response is, "Sure, all three of my lines are running with a Cpk above 1.3 today," I would say, "Meet you soon and you're buying!"

---

* Charles "Chuck" Knight (born 1936): CEO of Emerson Electric from 1973 to 2000, president from 1986 to 1988 and 1995 to 1997, and chairman from 1974 to 2004. Currently serves as chairman emeritus. In 2005, he wrote *Performance Without Compromise*, published by Harvard Business Review Press.
† Dr. Dale Besterfield, supra.

We'll discuss Cpk in Chapter 4, but the point is just these simple numbers based on SPC give me a picture of the situation at Mike's factory. It doesn't matter what his product is, or how many people are working for him, I can quickly tell if things are going well or not.

SPC is also a tool for monitoring and identifying process continuous improvement projects. Use it as such! For more on Dr. Deming, I suggest the books by Mary Walton.*

Since Dr. Deming was such a huge proponent for the use of SPC, let's explore his beliefs a little more in depth. Let's examine the Fourteen Points.

1. *Create a constancy of purpose for improvement of product and service*: This is an utterly brilliant sentence! While one could write a whole chapter on the philosophical brilliance of this point, this is not the place. Take a bit of time to just think about it, but be constant in your plans for improvement. Never stop improving. Never stop looking to the future. Is there any business that has senior management that believes they cannot improve? I'm sure there are. I'm sure that they have some of the Seven Deadly Diseases, too!

   This point could also be used as a starting point for risk-based thinking for ISO-9001:2015.

2. *Adopt a new philosophy*: Adopt a new philosophy? Well, if a business is not doing well, something should change! However, a successful business could adopt the new philosophy of the Fourteen Points and avoid the Seven Deadly Diseases. Let's look at a brilliant American company that no longer exists. Motorola was founded in 1928 and for decades came out with new and innovative products based on their incredible research and development. They were part of the backbone of American technological edge. Motorola invented the Six Sigma methodology, although General Electric (GE) was more successful in putting Six Sigma to practical use. Then, in 2007, Motorola started experiencing huge losses. They initiated massive reductions in their workforce, including their research and development divisions. Key executives left in great numbers. By 2011, Motorola no longer existed as an independent company. Did they fail to adopt a new philosophy? Did they fail to have a constancy of purpose?

3. *Cease dependence on mass inspection*: One of my absolute favorites! American and Western businesses continue to attempt to inspect quality into their product. If a lot fails the first time, just re-sample it or do a 100% inspection … another one of my sources of workplace humor … also known as sorting. People are not machines. People make mistakes. Just how accurate are people at inspecting products?

---

* Mary Walton (born 1941): Wrote *The Deming Management Method* in 1986, published by Putnam Books and *Deming Management at Work* in 1988, published by Perigee Books.

While I cannot cite the studies, I've read research that shows that overall people are about 85% accurate when conducting inspections. Some people are certainly better at inspecting and some people are worse. Regardless, human beings are never perfect. Does inspection accuracy decrease right before lunch, on a Monday morning, or on a Friday afternoon? Yes, it most likely does! As most quality assurance professionals know, the answer is to simply build quality into the product from the beginning. However, other management disciplines do not understand or allow this activity. "Just inspect it ... it's cheaper." Yeah ... right!

4. *End the practice of awarding business on price alone*: To this day materials managers and purchasing agents are rewarded for procuring a component or material at a lower price. They and most senior managers have no concept of the overall costs of poor quality. If you have low quality coming in your back door, you will produce and ship low quality, regardless of how much inspection and labor is put into the product. The quality of the component and material must be considered as well as on-time delivery, stability of the supplier, flexibility of the supplier, and other supplier attributes.

5. *Improve constantly and forever the system of production and service*: Very profound! I hope by now one can start to understand how the Fourteen Points are interrelated. Never, ever stop improving. Never, ever stop learning as an organization!

6. *Institute training*: Most managers in the United States and the Western world cannot see the value in training, whether it is formal classroom or on-the-job training. Rather than a value-added activity, they see training as an unnecessary cost in labor. This is especially true for the hourly workers. Taking a production worker away from making product to provide training is just a concept that these managers cannot grasp or accept.

   The benefits of a good training program are not always visible on a short-term basis. But, the value is there. A recommendation here is Dr. Phil Crosby's *Quality Is Free*,* to understand the hidden costs of quality that can be eliminated by a good training program.

7. *Institute leadership*: This one is the most abstract of the Fourteen Points. I think it is very difficult to just institute or train in leadership, although I know of some businesses that have had some moderate success doing this. My interpretation of this point is to hire leaders! If an organization finds a good leader, hire them whether there is an opening or not. Their talents will become extremely valuable over time.

* *Quality is Free*: Written in 1979 by Dr. Phillip (Phil) Crosby (1926–2001) and published by McGraw-Hill.

I've had some success in mentoring younger managers to be better leaders. But, if one does not have leadership ability, it cannot really be taught. This supports my point that good leaders are very valuable.

8. *Drive fear out of the workplace*: Does anyone want to come to work to be reprimanded or ridiculed? I am acquainted with a senior manager who was driving into work on a beautiful late spring day. The closer he got to work the more knotted up his stomach became and his stress increased. He wasn't even on the job yet. He had done nothing wrong. But, he worked for a nonviolent sociopathic CEO. That day, he notified the CEO that he was retiring. He just didn't need the stress any longer. This company lost an excellent senior leader. Fear is catastrophic to an organization!

Too many managers thrive on creating fear in their area of responsibility. Notice that I call these people managers and not leaders. A leader would never exhibit the behavior of creating fear in the workplace. A manager that creates fear in the workplace is an insecure manager. The organization is better off without them.

Consider the Seagull Manager, as coined by Ken Blanchard* in *Management of Organizational Behavior*: "A Seagull Manager will fly in, make a lot of noise, dump on everyone, then fly out." They only show up when there is a problem, contribute nothing, create fear, and are not leaders.

9. *Break down barriers between staff areas*: All departments should be working together for the common good of the organization. A leader will understand this. The average manager thinks only of their department. Institute organization-wide teamwork and cross-training.

One concept that was instituted at a company that won the United States Malcolm Baldrige National Quality Award was that a downstream person or department did not have to accept inferior or defective work from an upstream person or department. How profound!

The key point here is that all should be working to the common good and success of the organization.

10. *Eliminate slogans, exhortations, and targets for the workforce*: Why not? Most of these are just decorative wall hangings. Again, leadership comes in to play here. Remember that the Fourteen Points are interrelated. A good leader does not need slogans, exhortations, and targets. These are a waste of time and energy that the leader does not need.

---

* Kenneth "Ken" H. Blanchard (born 1939): Co-wrote *Management of Organizational Behavior* in 1968, and published by Prentice-Hall. In this book, he defined the term "Seagull Manager."

11. *Eliminate numerical quotas*: There are some businesses that pay work-ers by the piece produced rather than by the hour. The goal is more production. Quality of the product is a distant second. Eliminate long-term quotas such as sales quotas. It is just that simple. Someone is either doing a good job or they are not, regardless of their numbers.

What happens when a sales person exceeds their sales quota? I know of one instance where a relatively new salesman to an orga-nization more than doubled the sales of his predecessor and greatly exceeded his quota. The result was his boss telling him that he did great, but that he was not paying him the bonus he had earned, as it was too much money. The boss then raised his quota to exceed his current performance. This is a true example—so much for this man's enthusiasm.

The other thing that happens to those in sales is that when one continually meets or exceeds their quota, they are promoted to sales manager. So the best sales person is taken out of sales and put into management, where they may not be successful at all. Why not just pay the excellent sales person a higher salary as a reward?

Departmental production quotas just cause unnecessary stress and fear. These can also lead to cutting corners or eliminating prod-uct quality in order to meet the quota. If people are doing a good job, there are many ways to reward them. Use rewards as part of leader-ship rather than numerical targets.

12. *Remove barriers to pride of workmanship*: This one is also somewhat philosophical and yet brilliant. Consider that people want to do a good job. They also usually know where the problems are, even if they have some trepidation or don't understand the right terminol-ogy. They are often starved to get someone to listen to them. As a consultant, I have had great success using this principle. I just ask the hourly worker what the problems are. Most will give honest answers and most point to bad, inattentive, or other types of management that we've discussed. I simply take the information from the worker, organize it in a pragmatic fashion that suits the organization and I'm a hero. The answers were there all along, but there were barriers.

People want to do a good job. This crosses cultural and religious boundaries. Leaders will give them the opportunity to do so.

13. *Institute a vigorous program of education and retraining*: Tom Peters* has stated, "If your company is making money, double your train-ing budget. If your company is losing money, quadruple it." A very

---

* Thomas "Tom" J. Peters (born 1942): American writer on business management practices known for his drive for excellence.

profound statement from another brilliant guru. The more typical process when a company is losing money is to cut the staff.

There are some organizations that rely heavily on written procedures. This may be driven by a misunderstanding of industry regulatory requirements. Their training program will often consist solely of a review of the procedures every year with documentation of the review. Done! Was anything actually learned? Was there value in this process?

At one time, I was employed by a company that had instituted skill-based training for all hourly employees. Most of the classes involved in acquiring a skill required the passing of an exam or quiz. As participants succeeded in passing classes, they received a raise in pay. Some jobs required the passing of designated classes to secure a permanent job. Then, to move up to the next grade level, again required the passing of designated classes. Some workers were content to stay at a lower grade, but most wanted a higher wage and therefore strived to move up in grade. Not all classes were directly related to the job. Many were more general classes about business, quality, and leadership. The result was a well-paid, happy, productive workforce that produced a very good product.

14. *Take action to accomplish the transformation*: Just do it! Hint: It will take leaders!

Another example of an old set of principles that are still very relevant today was developed by Henri Fayol.* Henri Fayol was a French engineer that developed a business management theory known today as Fayolism. He published *Administration Industrielle et Generale* in 1916. Despite World War I, the demand for his book was immediate and by 1925, 15,000 copies had been printed. The first English edition was in 1929 in Great Britain. Despite widespread interest, it was not published in the United States during that era. His work has been recently republished in English for the United States by Martino Publishing in 2013 as *General and Industrial Management*.

I find it most interesting that he also developed fourteen principles of business management.

1. Division of work

2. Authority

3. Discipline

---

* Henri Fayol (1841–1925): A French mining engineer who wrote *Administration Industrielle et Generale* in 1916. His theories on scientific management are known as Fayolism. He is considered one of the founders of modern management methods.

4. Unity of command

5. Unity of direction

6. Subordination of individual interests to the general interests

7. Remuneration

8. Centralization

9. Scalar chain (line of authority)

10. Order

11. Equity

12. Stability of tenure of personnel

13. Initiative

14. Esprit de corps

Clearly, this is a book on statistical process control with a focus on a unique instruction method and the introduction of the concept of "problem variation." However, SPC is used to monitor, control, and measure improvement in a production or service process. That is, improving quality, which improves the business. This brief review of Fayolism is to provide potential interest to further explore this exceptional work from over 100 years ago and apply it today to the reader's business or organization.

Other suggested reading is *Principles of Scientific Management* by Frederick Winslow Taylor,* originally published in 1911 and recently republished in 2007 by NuVision Publications, LLC. Taylorism, as Frederick Taylor's principles are known, is rather outdated today, but it is an interesting look back into beliefs of the early twentieth century. Then definitely read *Quality is Free* by Philip B. Crosby† in 1979 which outlined his 14 steps (again) to quality improvement, as published by McGraw-Hill Book Company. It is a timeless classic.

---

* Frederick Winslow Taylor (1856–1915): An American mechanical engineer, who wrote *Principles of Scientific Management* in 1911, published by Harper and Brothers Publishers. He is known as the father of the scientific management and efficiency movement.
† *Quality is Free*, supra.

QUALITY IS REMEMBERED: LONG AFTER THE PRICE IS FORGOTTEN

John Ruskin* (1819–1900) said, "It is unwise to pay too much; but it is unwise to pay too little. When you pay too much, you lose a little money ... that is all. When you pay too little, you sometimes lose everything ... If you deal with the lowest bidder, it is well to add something for the risk you run; and if you do that, you will have enough to pay for something better."

---

* John Ruskin (1819–1900): An English writer, artist, art critic, draftsman, social thinker, and philanthropist. He wrote many essays and treatises, which has led him to be quoted to this day.

3

A Teaching Methodology That Works

Perfection is achieved, not when there is nothing more to add, but when there is nothing left to take away.

Antoine de Saint-Exupery

A leader teaches with patience. A manager without patience is no leader.

Rafael Aguayo

Wisely and slowly; they stumble that run fast.

William Shakespeare
Romeo and Juliet

Introduction

When one is scheduled to take a training class in Statistical Process Control (SPC), it is often viewed as a less than desirable way to spend one's time. Many employees have less than pleasant memories of activities in a classroom setting. Others may fear the fact that it is a technical subject that they may not readily understand. Even those with a basic understanding of SPC may dread the class because, let's face it, statistics is not an exciting subject.

Who Are the Audience and What Is the Time Involved?

The success of a class, at any level, is largely dependent on the instructor and the instructor's preparation. As one who has been called upon many times to share my expertise in SPC, I have developed a methodology for providing a basic understanding of SPC in a fun and enjoyable setting ... at least to the extent allowed by the subject matter and workplace decorum.

The methodology presented in this chapter has been successfully used to instruct people from a wide variety of skill and knowledge levels. Participants have ranged from hourly workers that were uncertain as to how to use a calculator to senior managers with at least some education in science and mathematics. As the instructor, I found the most challenging were the classes consisting mostly to entirely of engineers, as they wanted to indirectly teach the class. The most enjoyable part for me were the classes with hourly workers with a strong desire to learn what, for them, was a difficult subject.

Three days is the best time to adequately present this material in its entirety. If the participant skill level is high, then it is possible to complete the class in two days, but it forces the instructor to move quickly with less discussion. The most time that I've taken to present this class is four days. Another forum I've used is to present the class in two days just covering the basics of SPC, excluding the more complex portions included toward the end of this chapter.

If possible, it is advantageous for the instructor to determine the skill level of the participants ahead of time, ideally by contacting the participants and asking a few questions on their background. This is easily accomplished by email. I have taught this class many times where the number of days involved was predetermined and I had minimal information on the participants when I walked in the classroom door. However, in this situation I always have the participants provide, as part of their personal introduction to the rest of the class, a brief summary of their background as related to SPC or mathematics. These introductions are normally part of every class and should take from 30 seconds to no more than a minute.

Regardless, this class is best presented over three full days. It can be as long as four days. However, some organizations only allow two days. This class cannot be adequately taught in less than two days. However, if the class is a mixture of various backgrounds and there is a choice, then the instructor will have to, as much as possible proceed at a rate that fits those with the least applicable background.

Then consider this material on a college level where class time may only be an hour per session, two or three sessions per week. The instructor will have to do their best to determine the start and stop points as this material was developed for a multi-day seminar format.

Before beginning the class in SPC, provide the items listed in Figure 3.1.

Note: The best class size range is from five to 15 people. With larger groups, it is not necessary to obtain answers on each activity from each participant as described in this chapter. However, each student should still participate on an alternating basis. In addition, plan on a break at approximately every two hours in addition to lunch. It is important to provide these breaks to enhance the learning experience and retention of information. If possible, also have refreshments available in the classroom.

Materials list

Nine blank sheets copy paper/trainee (have extra sheets of paper available)

One yellow highlighter for all participants

One other color highlighter for about 25% of the participants

One 12" ruler/trainee (do not provide until later in the class)

One notepad/trainee

Pens and/or pencils

One handout of class material/trainee

Calculators (optional)

Other related training aids

Refreshments (bottled water, soda, ice and cups, etc.) and snacks as possible

Materials list for instructor

Calculator

Dry Erase Board(s) – large size

Pen or pencil

Multiple colors of dry erase markers and eraser

FIGURE 3.1
Materials list.

Part 1: Introduction

Start the class with a general discussion of product quality. Ask what it means to each participant. List their ideas on the dry erase board. This helps the instructor to also gauge the level of knowledge of the participants. Answers can and will usually vary greatly. If the class resists participation, introduce the concept that everyone is a consumer. Everyone will have an opinion on the quality of products used in their daily lives: foods, appliances, clothing, and vehicles. Therefore, each participant should at least have a good idea of what poor quality is. If the class is presented to participants from business, bring the discussion of these everyday items to the types of service provided or products manufactured by the group(s) in the class. The concept that product quality should focus on the customer can be illustrated with this type of discussion. In addition, linking the discussion to products the participants are involved with helps to engage the participant and translate the topics into their everyday work, which makes the learning more applicable.

The concept that each customer is different will help to further define quality. Conduct the following survey and list the results. Ask each class participant what make and model of vehicle they normally drive (e.g., Ford Mustang). Follow with the question, "When that vehicle was new, right off the assembly line, was it a good quality vehicle?" Most people answer, "Yes, it was." Many different makes and models of vehicles will be listed and, since most will agree that theirs was a good quality vehicle when new, it will be relatively simple to illustrate the importance of customer perception of good quality. About one participant in 10 will try to start a debate on the quality of their vehicle and control the discussion. Just acknowledge their comment and move on. It is important to keep the discussion moving and focused.

Connect this information, along with the instructor's knowledge, into a definition of quality. The instructor should have the class help to develop the definition. Compare the developed definition with various published definitions of quality, such as from Dr. Juran or Dr. Fiegenbaum.

Dr. Juran[*] defined quality in 1951 in his *Quality Control Handbook*, published by McGraw-Hill, as, "Quality means that a product meets customer needs leading to customer satisfaction."

Dr. Fiegenbaum[†] stated in a 2014 interview with *Industry Week* magazine that, "Quality is what the user, the customer says it is."

My personal definition of quality is: "Quality is the perception of what the customer thinks they need, want, and desire!"

This exercise requires all to participate. It is a good tool to get everyone involved at the beginning of the class. The participants should now be more relaxed, engaged, and focused on quality. They are now ready to move on to SPC.

Part 2: Simulated Factory

Create a "mini-factory" to make a good quality product as defined by the class. Each participant needs to identify their nine sheets of copy paper, in order, as follows; 1A, 2A, 3A, then 1B, 2B, 3B, and then 1C, 2C, and 3C. Each participant will have three groups of three sheets, an "A," a "B," and a "C"

---

[*] Dr. Joseph "Joe" M. Juran (1904–2008): An engineer and management consultant wrote the *Quality Control Handbook* in 1951, which was published by McGraw-Hill. The several later editions of the handbook were co-written with others.

[†] Dr. Armand V. Feigenbaum (1922–2014): An American quality control expert and businessman whose concepts helped inspire total quality management.

group. This order must be maintained. As easy as this is, some will have trouble following these simple directions. Have extra sheets of blank paper available, as needed. Be patient. Help as needed.

On the dry erase board, identify the specification for manufacturing the product as a four-inch by six-inch upper case letter A as shown in Figure 3.2.

Note that a tolerance is not provided. One piece of paper represents one unit of product, with one capital letter "A," to be made on each piece of paper. The instructor should remind the class not to start until instructed to do so since production does not start until the beginning of the shift in the real world. Draw some examples of what the product is not (e.g., a tiny "A" that is nowhere near four-inch by six-inch, more than one "A" per page, or a very large "A", etc.). See Figure 3.3.

FIGURE 3.2

FIGURE 3.3

Only three "A"s are made in each simulated shift. First the A group, then the B group, and then the C group. The first "A" to be manufactured is on sheet 1A, the second on 2A, and the third on 3A. The "A"s are to be manufactured using the highlighter marker provided. Ensure that part (about 25%) of the class is not using yellow highlighters, but the other color provided. Allow 60 seconds to make the three "A"s and when ready have the class start their production shift.

As time passes, tell the class participants where they are in their production day. At approximately 15 seconds, tell them they just finished their first break. At approximately 30 seconds, tell them they just finished lunch, then at approximately 45 seconds, they have finished their afternoon break. The instructor may point out obvious defects while the class is in production by saying for example, "The QC [quality control] inspector may reject that one." Keep the activity light and fun, but do not criticize. After 60 seconds, if participants are not finished, announce that they are now on mandatory overtime.

After each participant has completed their three "A"s, have the class evaluate their production day's output. Discuss defects, scrap, equipment problems (marker or paper), operator errors, lack of communication from management, and so on. These are regular issues that we face in manufacturing, which can be illustrated in this activity. Also, discuss efficiency if there was excessive overtime; however, it is imperative that the instructor not allow any ridicule of a participant's product!

Next, the instructor illustrates a real world problem by stating, "Hey wait a minute! We just received a call from our customer. They ordered *yellow* "A"s and want to know why they received some *green* (or whatever other color of highlighter was used)?" Use this illustration to define an *attribute defect*. Discuss other types of attribute defects. A soda can is a good prop to use since you should have class refreshments. Discuss possible attribute defects that can occur on a soda can. Discuss some attribute defects manufactured by the company the participants work for. Use other examples that you may have.

Ensure that all participants now have only yellow highlighters.

Now that the class participants are trained and experienced "A" producers, get ready to produce the second set of "A"s on sheets 1B, 2B, and 3B. New manufacturing instructions to be given to the class are to use only yellow highlighters and production time has been cut to 40 seconds. Follow a similar scenario as with the first group of three "A"s, noting break times at approximately 10-second intervals. The instructor should remind the class of the importance of quality by stating, "Now class, remember, our customer is expecting quality to improve!" This is a good example of continuous improvement. If time allows, continuous improvement concepts can be illustrated throughout this exercise. The instructor should announce to the class statements such as, "Now that you are experienced, management is expecting you to be more efficient." Discuss the concept of improved efficiency through experience.

After completion, discuss the same manufacturing parameters, as occurred after the first group and compare the results. The instructor should ask questions to the class, such as:

- "Have things improved after training and experience?"
- "How is product quality?"
- "Is the class meeting specifications?"
- "Was overtime still needed?"

Discuss obvious defects and manufacturing problems. Again, keep the discussion light and fun, while not allowing any ridicule.

Prepare the class to manufacture "A"s with the last three sheets of paper, the "C" group. The instructor should announce, "Management has invested in some capital equipment to improve the quality of our product." Hand out a 12-inch ruler to each participant. Remind the class that the specification for an "A" is six inches high. Note to the instructor that the specification is not the length of the leg of the "A"; rather, it is the total height. Some participants will use the ruler to make the legs of the "A" six inches long. Allow the class to have 60 seconds to complete the C group of "A"s. Avoid making any statements regarding the four-inch width specification. If asked, just ignore the question and tell the class they are on the clock and the production day is ready to start. Does this simulate the attitude and behavior of many line managers? Discuss this at the end of this simulated production day. Again, let the class know where they are at in their simulated production day.

Upon completion, discuss the quality of the product again. Ask the class the following questions:

- "Did the equipment (i.e., the ruler) improvement help improve product quality?"
- "How is product quality?"
- "What problems were encountered?"
- "Did management provide adequate training to use the new equipment?"

Discuss how the issues encountered in the class relate to what happens in the real world of manufacturing. Lack of communication, equipment problems, inadequate training, and operator error are some of the problems that can be illustrated in this exercise. For most participants, using the ruler will take more time. Some will use the ruler to draw straight lines while making their "A"s as well as attempting to meet specifications. Remember, you, as management, gave them virtually no training on using the new equipment. Does this represent the real-life manufacturing world in many organizations? What kind of problems did the lack of training cause? Typically, introducing the use of the ruler causes the participants to take more time, since they try to be perfect. Nothing was said about straight lines, just to draw a six-by-four-inch "A." However, virtually all will use the ruler as a straight edge, which will take more time. People really want to do a good job and so will naturally try to make straight "A"s as their product. Problems arise from providing workers with new equipment without training or communication:

- Straight lines on the "A." This is known as over production in a Lean Manufacturing sense; that is, features that the customer does not need or want.
- Striving for perfection. The perfectly sized "A" with straight lines.
- Workers trying to figure out what to do with the new piece of equipment.
- Time wasted.
- Workers discussing among themselves what to do.

All of these issues are the fault of management. As a manager, one should remember this!

If the instructor has the knowledge, this is a good time to discuss some of the classic wastes as defined in Lean Manufacturing. This is also a good time to relate the produced product back to Dr. Deming's Fourteen Points. Again, this is a class to provide a foundation to improve quality. If time allows, these extra minor discussions about Lean Manufacturing and Dr. Deming are very useful to those attending the class.

The instructor should provide an update to the class at this time by stating, "Well guess what? Our customer is still not satisfied! Yes, we are making yellow 'A's that are approximately four inches by six inches, but we left out a critical dimension. Our customer specified yellow 'A's with a crossbar of two inches plus or minus a quarter inch!"

Have we met that specification? Why was this critical dimension not communicated to manufacturing personnel? Why did management or engineering not tell the production workers? Obviously, we are not sure. Does this type of communication lapse ever happen in manufacturing? Sure it does, especially in multi-site organizations. Spend some time discussing this lapse of communication from production management or engineering.

- Managers are often too busy. Often their work is full of bureaucratic waste. Again, if possible, think of Lean Manufacturing.
- Managers sometimes spend their time trying to please their boss. They are focused on getting a promotion, rather than the job at hand.
- Sometimes engineers really do change the customer's specification. They think that if the product is produced as they design it, the customer's specification will be met.
- Some people in leadership-type roles are just not good at communicating. There are, frankly, just a lot of really bad managers.

This is not a management-bashing class. I'm merely pointing out that virtually all problems have a root cause that ultimately belongs to management. Managers are paid to lead, so lead they must. If a piece of equipment breaks down, then a manager made a decision not to maintain it or replace it. It is not the machine's fault or the workers.

Part 3: Plotting Class Data

The instructor must now draw a matrix on the dry erase board as shown in Figure 3.4.

	1A	2A	3A	1B	2B	3B	1C	2C	3C

FIGURE 3.4
Matrix drawing.

The instructor should start this section by stating, "So class, any idea how we determine how well we are meeting the customer requirement of yellow 'A's with a crossbar of two inches plus or minus a quarter inch? Sure, we can measure! Do we need to measure each 'A'? Will it be easier to sample and just measure some of the 'A's?" Of course it will, and in manufacturing, time is money. Explain to the class that you are going to illustrate the basic concepts of SPC. Again, this learning process works better with the larger class size. Ensure the class understands that SPC is not used in 100% inspections. Remember point number three by Dr. Deming. It is used to evaluate production by sampling a small portion of the overall output. 100% inspection is more appropriately called sorting.

Using the information for manufacturing the "A"s, define the terms *target*, *specification*, and *tolerance*. Also, define *upper specification limit* (USL) and *lower specification limit* (LSL). Discuss how the specification or tolerance is what an engineer believes the product should be built to. Note that in reality a design engineer's desired specification may not be able to be met. Although often the specification from the customer's engineer must be adhered to!

Before it can be determined how well the product is meeting the customer's specification, we need some data. Explain that the class is going to measure the length of the crossbar to the nearest eighth of an inch, starting with 1A and proceeding in order through to 3C.

It is important to note a couple of items. The width of the leg of an "A" will be approximately an eighth to a quarter inches wide, depending on the type of marker used and the student's method. This uses up the tolerance provided. The correct measure is just the crossbar inside the legs of the "A." Expand this concept to demonstrate how operator measurement error can make acceptable product appear to be defective (*Type I* or *Alpha Error*) or defective product can appear to be acceptable (*Type II* or *Beta Error*).

Of course, if the "A" is sloppy and the crossbar extends past the legs of the "A," then the full length needs to be measured. If needed, take the time to show how to use the ruler. Yes, there are people that are unsure of how to use a ruler. This will avoid embarrassing any of the class participants. Define the terms *subgroup* and *lot* or *batch*. Briefly discuss with the class how a subgroup of five works best with SPC. While the mathematics will not be covered, explain that in statistics a subgroup of five works best. Explain that later in the class it will become more apparent as to why a subgroup of five pieces of data is best. Ask the class to just accept this point.

There will be a lot of variance in the measurements. Measurements should be rounded to the nearest eighth of an inch, which will make it easier for the class to measure. Unless of course the participants are engineers! But, it will also be easier for the instructor who will need to quickly convert the fractions to decimals for recording on the dry erase board. It helps if the instructor is good at converting fractions to decimals in their head very quickly. If not, prepare or use a decimal equivalent chart. Have the first participant give their measurement of the crossbar on their product number 1A. Convert to decimal (e.g., $2 \frac{1}{8} = 2.125$) and fill in the measurement in the first box under 1A in the matrix (Figure 3.4). Follow with the next four participants with their measure of 1A, converting to decimals and filling in the matrix accordingly. The sixth participant will then provide the measure of their product number 2A. Record on the matrix and follow with the next four participants. If the class size is small, the participants will provide measurements in multiple groups of "A"s. However, it is a better class if all nine "A"s of each participant are not recorded. This is why a class size of more than five participants is preferred. Again, do not allow any ridicule!

Continue measuring a subgroup of five As in each of the remaining product groups and recording the measurements in the matrix. When finished, circle in red the out-of-specification conditions. (The acceptable tolerance is 1.75–2.25 inches.) Look at each subgroup. Ask the class, "Should the subgroup and associated production lot or batch be rejected?" Their answers should vary. Discuss that to reach a decision, additional information is needed.

At this point, a discussion of variance is appropriate. Define the terms *variance* and *variable defect*. Demonstrate using the information in the matrix. Define the parts to a process: materials, methods, machinery, measurement, manpower, and environment as per Ishikawa, as we discussed in Chapter 1. Discuss how all of the variances involved in the individual process parts make up the total process variance:

- None of the parts to the process are perfect.
- Are the materials used to make your product perfect every time?
- What about the equipment used to make your product?
- Well, people certainly are not perfect!
- Everything has variation.
- In a process, all of the different parts have variance, which contribute to the total process variance.

Two people will drive the same car differently. Two cars of the same make, model, and year will not perform exactly the same. This is variance. Discuss this topic in some detail. Define *common cause variation* and *assignable or special cause variation*. Give examples of each.

- Data entry error or measurement error is assignable cause.
- Machine break down is assignable cause.
- An operator adjusting a machine is an assignable cause ... although the adjustment may possibly be necessary.
- Lack of operator training is an assignable cause.

- A machine that malfunctions, but corrects itself is an assignable cause if it happens infrequently.
- Picturing a manufacturing line with multiple machines, the cumulative lack of perfection of each machine is common cause.
- The lack of perfection in a specific machine is common cause.
- In a manual labor situation, the lack of perfection among the personnel involved is common cause.
- Measurements that are not overly exact, but yet acceptable are common cause. For example, measurements are taken in whole numbers, when measurements to two or three decimal points would be more accurate for the activity involved.

These can be difficult concepts for some to grasp. A lot or little time can be spent on defining variation, depending on the class time parameters and how well the participants grasp the concept. Correlate the discussion back to the earlier discussion of cars to define quality. What is important is that the class participants understand that variation is the root of all quality problems. While ultimately management is responsible, the variation that exists or that they cause is the root of the problem. As control charts are developed later in the class, common cause and assignable cause variation will be more apparent.

In the third square from the bottom in the left column, add "Xbar" or subgroup average to the matrix as shown in Figure 3.5.

	1A	2A	3A	1B	2B	3B	1C	2C	3C
Subgroup average									

FIGURE 3.5

Calculate the average (or mean) of each subgroup and record these values in the matrix. Define the term *Xbar* or *subgroup average*. Circle in red the out-of-specification subgroup averages. Some of the subgroups will fall within the two plus or minus a quarter inch tolerance, even though some of the individual measurements are outside of the tolerance. Ask the group if the subgroup average is a good measure of quality. Discuss various ideas about the variance of the class measurements. Note that each of the five measurements recorded in a subgroup can be out of specification, but the subgroup average can be within specification. Close the discussion with the determination that additional information is needed.

In the second square from the bottom on the left side, add range (R) to the matrix as shown in Figure 3.6.

Define the term *range*. Determine the range between the high and low measures of each subgroup, and list in the matrix. Circle in red those ranges greater than half an inch (the tolerance is plus or minus a quarter inch or half inch).

Now evaluate the subgroups for acceptability based on the Xbar and R information. Discuss which groups of product; the A group, the B group, or the C group should be shipped to the customer. Based on this information, which subgroups represent product lots that should not be shipped to the customer and require some type of correction. Discuss that the five measurements recorded on the matrix can represent hundreds of pieces of production. Bring in the concept of how random statistical sampling can provide an accurate picture of the quality of the entire population. Point out how this can save time in manufacturing and is as equally effective as inspecting each

	1A	2A	3A	1B	2B	3B	1C	2C	3C
Subgroup average									
Range									

FIGURE 3.6

	1A	2A	3A	1B	2B	3B	1C	2C	3C
Subgroup average									
Range									
Process average									
Range average									

FIGURE 3.7

or most pieces of product. Present to the class that 100% inspection or sorting is simply trying to inspect quality into the product.

To the matrix, add the term process average and range average to the lower left squares as shown in Figure 3.7.

Define the term *process average* as the average of the subgroup averages. It is the average of the process. It is also called the *grand average* or *Xdbar* or *X double bar*. Define the term *range average* or *Rbar*. Calculate the process average and range average for the class and add to the matrix.

Now we have some information about product performance. But the product in each of the subgroups is history. Ask the class, "What can we do while the product is being manufactured?" Do not erase the matrix of data if at all possible. If you must, then copy the matrix with data on a sheet of paper. The data will be needed later in the class.

Part 4: Bell Curves and Normalcy

On a dry erase board draw a "normal bell curve" as shown in Figure 3.8. Define a *normal bell curve*.

Discuss how it is derived. It is important that the class understand that the normal bell curve is mathematically real and not just a theory used to teach statistics. I have found that using the analogy of the grading curve used in school is a good way to explain this concept. Almost everyone has a story

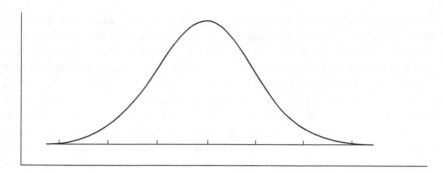

FIGURE 3.8
Normal bell curve.

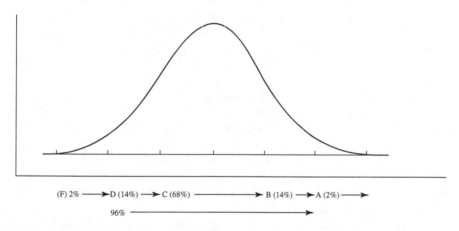

(F) 2% ⟶ D (14%) ⟶ C (68%) ⟶ B (14%) ⟶ A (2%) ⟶

96% ⟶

FIGURE 3.9
Normal bell curve divided into six equal parts.

of getting 90% correct on a test, but getting a grade of C (or 40% correct and getting an A). Divide the normal bell curve into the standard six equal parts as per Figure 3.9. Label for the letter grades and percentages as shown in Figure 3.9.

It is important to point out that the middle two sections (zones) comprise the C grades. Explain how the amount of Fs equal the number of As, the number of Ds equal the number of Bs, and the Cs have the greatest area of any grades. This is a way to demonstrate that the area under the curve represents the relative amount of data at that point. There are a lot of Cs but very few As and Fs. Label the Cs as comprising 68% of the total number of grades (each individual C zone comprises 34% of the total grades). Label the next two zones, which are the B and D grades, as 14% of the total for each grade classification. Finally, label the remaining zones, which are the A and F grades, as each having 2% of the total. Label the middle four zones together as comprising 96% of the total grades. Another way to look at it is 96% of the population (those receiving the grade) receive a B, C, or D. It is important to understand that each bell curve can only represent one type of product or population. If one desires to plot the grades from three different classes, then one will need three bell curves.

Re-draw your normal bell curve without the grade information or percentages and replace with the more specific percentages as shown in Figure 3.10.

Now discuss the six zones of the normal bell curve, as it relates to quality defects and the more precise percentages are shown. Point out that the total does not equal 100%; rather, the total under the normal bell curve is 99.7% of the total population. Point out how the curve extends just slightly past the hash marks on each end which account for the remaining 0.3% or 0.15% on each end. Statistics are a great tool, but not 100% perfect. It is important to remind the class that these values or the area under the curve are calculated. The mathematics involved is not appropriate for a class at this level.

Explain how the midpoint of the curve or the center of the line is the process average or X double bar or grand average. Emphasize that the process average is always the center of the curve.

The concept of standard deviation or sigma units is the next topic. Re-draw the normal bell curve as shown in Figure 3.11.

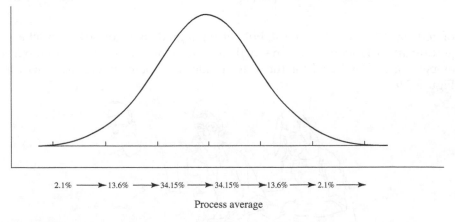

2.1% ——▶ 13.6% ——▶ 34.15% ——▶ 34.15% ——▶13.6% ——▶ 2.1% ——▶

Process average

FIGURE 3.10
Normal bell curve with more specific percentages.

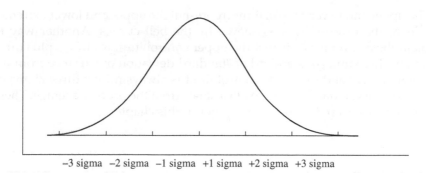

-3 sigma   -2 sigma   -1 sigma   +1 sigma   +2 sigma   +3 sigma

FIGURE 3.11
Normal bell curve with sigma units.

Explain that the zone limits are the units of the amount of variance from the process average (or the average grade). Each section or zone of variance is a *standard unit of deviation* from the process average or a *sigma unit*. Explain how when one moves to the right of the midpoint or process average the limits are labeled plus one sigma, plus two sigma, and plus three sigma. Then, as one moves to the left of the process average, the limits are labeled minus one sigma, minus two sigma, and minus three sigma. Again, explain to the class that a sigma unit equals one standard unit of deviation of variance. Explain that the sigma units are always equal, that is, equal units of variation. Then remind the class that the area under the curve represents data and the data is measured from the process. At this time do not spend too much time on sigma units or standard deviation. We will cover this in more detail later in this chapter. Just focus on there always being three equal sigma units or three equal units of standard deviation of variance on each side of the process average or process grand average. Tell the class that these sigma units have nothing to do with Six Sigma methodology.

Define *upper control limit* (UCL) and *lower control limit* (LCL). Relate these terms to the bell curve as shown in Figure 3.12.

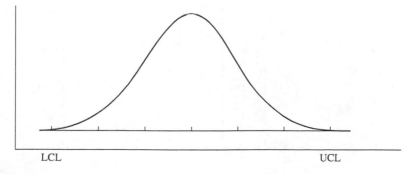

LCL   UCL

FIGURE 3.12
Normal bell curve with LCL and UCL.

The upper and lower control limits represent the upper and lower extremes of where the process lives as shown by the bell curves. Another way to explain the control limits is that the upper control limit is always plus three sigma or three units of standard of standard deviation of variance from the process average and the lower control limit is always minus three sigma or three units of standard deviation of variance from the process average. These terms will be covered in more detail later in this chapter.

Part 5: Process Data

The instructor should draw a sample histogram on the dry erase board to illustrate the concept of how the bars represents approximate amounts of data as shown in Figure 3.13. Define a *histogram*.

A histogram should now be developed using the class data from producing the As. Label the upper and lower specification limits. Overlay an approximate bell curve to illustrate the relationship to the histogram or data. Draw the curve as close to normal as the data will allow. It is important that the class participants grasp that the area under the bell curve at a given point, represents an amount of data at that point. Also, draw and describe different shapes of histograms that can occur in the real world: skewed left, skewed right, bimodal, flat, very centered, and so on. Some examples are shown in Figure 3.14.

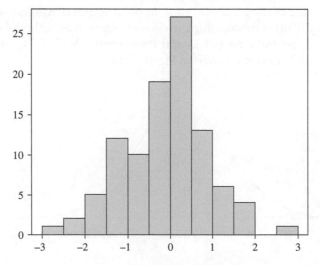

FIGURE 3.13
Sample histogram.

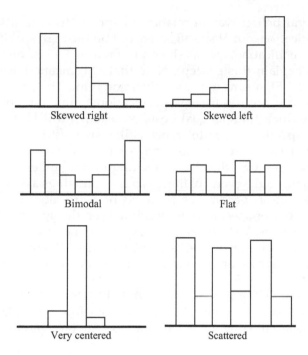

FIGURE 3.14
Different shaped histograms.

The fact that the bell curve is a graphic representation of the process is a very important concept for the class to grasp. Now draw an approximate curve over each of the types of histograms. The only curve that is "normal" is the very centered curve. The others are not "normal" curves. There are some causes for this:

- Not enough data (most common)
- An out-of-control process
- Process has shifted, and
- Assignable cause variation

In general, as one accumulates more data or measurements of the process, there is a tendency to approach normalcy. This is as long as the variation remains common and there are no (or minimal) assignable cause events. A bimodal distribution can be the result of an assignable cause event. Perhaps a machine was overhauled and now operates at a different point in the process. Or a bimodal distribution can represent the difference in how two different shifts or people run the same process. A scattered distribution is usually either a lack of data or an out-of-control process. A skewed distribution could be the natural distribution for that particular process.

Draw several bell curves in relation to specification limits, including within specification, out of specification, out on the high specification, out on the low specification, etc., as shown in Figures 3.15 through 3.19. This will help to explain this concept. Note that LSL means lower specification limit and USL means upper specification limit. Then ensure the class understands you are discussing specification limits, not control limits. Consider the thick lines as goal posts on a football field. Consider that meeting the specification or tolerance is like successfully kicking a field goal. The instructor can draw as many or few as class time allows and depending on how well the class is picking up the understanding of the concepts. Go through several of these types of drawings asking the class if there are any out-of-specification products being produced in each drawing. Ask if the out-of-specification condition is on the upper, lower, or both ends of the process. Ask the class to explain what is occurring in each drawing. Ensure that everyone is participating. All should understand that in these scenarios, all products are not meeting specification, except in Figure 3.19.

"In this example, the process as illustrated by the bell curve is slightly out of specification on the low side or using our football analogy, some of the field goals are missing to the left, but most are between the goal posts" (Figure 3.15).

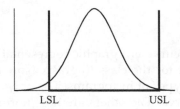

FIGURE 3.15
Out of specification, low side.

"Here we have just the opposite. Some of the process is out of specification to the upper side or some of the field goals are missing to the right, but most are between the goal posts" (Figure 3.16).

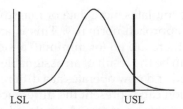

FIGURE 3.16
Out of specification, high side.

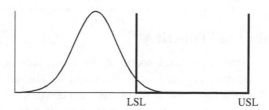

FIGURE 3.17
Almost all out of specification, low side.

"In this example, almost everything is out of specification with just a small amount of product that is in specification at the lower end of the tolerance. Virtually all of the field goals are missing to the left side" (Figure 3.17).

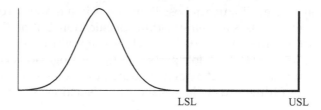

FIGURE 3.18
All out of specification, low side.

"Here we have an example of all product produced being out of specification or out of the tolerance range. All product produced is at the outside the specification on the low side, or all field goals are wide to the left" (Figure 3.18).

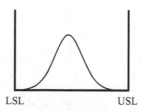

FIGURE 3.19
All in specification.

"In this example (Figure 3.19), all product is within specification with room for error on both the upper and lower sides. The process is centered. All field goals kicked are easily scoring points."

A lot of time can be spent on bell curves and graphic representations of a process or just a little, depending on the time scope of the class.

Part 6: Taguchi's Loss Function

Depending on the limitations and scope of the class, this is a good time to describe and discuss the Taguchi Loss Function as shown in Figure 3.20. Genichi Taguchi* was a Japanese statistician who developed a theorem that demonstrated that as a process approaches the specification limits, the more dissatisfaction, problems, or losses are encountered. In Figure 3.20, Taguchi's Loss Function is used to demonstrate dissatisfaction as shown in the dark gray square. The curved line represents the process. The upper and lower specifications are identified by dotted lines. The light gray shaded area represents problems, dissatisfaction, or losses. Taguchi's theorem states that as the process approaches the specification limits, there is an increase in losses and dissatisfaction. Think of the goal post analogy as previously presented.

Another way to view Taguchi's Loss Function is in a manufacturing environment. Consider parts X as being within specification, but on the extreme upper end. Part X is mated to parts Z, which is also within specification, but on the extreme lower end. Both lots parts pass QC testing. However, some of the parts X won't fit to parts Z as demonstrated by the Loss Function. This issue of parts not fitting leads to losses and dissatisfaction.

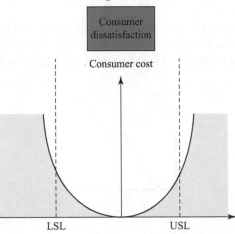

Part VI – Taguchi's loss function

FIGURE 3.20
Taguchi's Loss Function.

---

* Genichi Taguchi (1924–2012): A Japanese statistician and engineer who applied statistics to improve the quality of manufactured goods. His contributions include three principal areas: specific loss function, off-line quality control, and innovations in the design of experiments.

Part 7: Control Charts

Figure 3.21 shows a perfect process *control chart* template with a bell curve included. This is for demonstration purposes only and does not represent the real world. It is to show the relationship of the bell curve and histogram of data as it relates to the control chart.

Define in terms of process control. Explain that the information on the control chart will represent real data, just as the bell curve and histogram do.

The instructor can calculate the UCL and LCL for the data generated in the class. Just like the bell curve, a control chart can only have one type of data. If an organization is going to measure five variable items on their product, there will be five control charts. Since the calculations are not that difficult, it is better to have the class do the calculations. Or if the class participants are not technically qualified, the calculation can be demonstrated. Remember that the Process Average = the Grand Average = X double bar = Xdbar.

$$UCL = Xdbar + (A_2 Rbar)$$

$$LCL = Xdbar - (A_2 Rbar)$$

Values of A_2 at certain values of n, where n = the subgroup size:

n	A_2
2	1.880
3	1.023
4	0.729
5	0.577
6	0.483
7	0.419
8	0.373
9	0.337
10	0.308

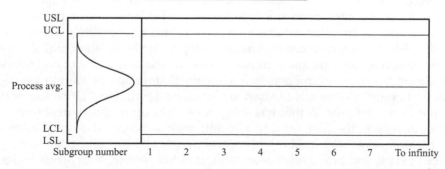

FIGURE 3.21
A perfect process control chart with a bell curve included.

As stated earlier, a subgroup size of five units is mathematically best. A subgroup size of six is the next best. If there is a choice, just choose a sub-group size of five.

If sigma or the standard deviation is known then the formula is as follows:

$$UCL = Xbar + 3\ sigma$$

$$LCL = Xbar - 3\ sigma$$

However, at this point in the class, sigma or the standard deviation will not be known.

Add the "mini-factory" specification limits, target, process average, and control limits to the control chart on the dry erase board. The control limits will probably have to be moved, since the class data is not likely to fit within the specification limits. Add the subgroup averages from the products mea-sured from the production of the "A"s to the Xbar control chart on the dry erase board. Note that with a large class, one can record more subgroups on your original matrix. This will provide for a better control chart, but the nine subgroups normally plotted on the control chart are sufficient to dem-onstrate the concepts.

Now, with the class's help, plot each of the subgroup averages on the con-trol chart and connect with lines in the order of the subgroups. Also, add the process average or grand average to the control chart. There is almost always an out-of-specification condition. "So class, what about this point here? Does the chart indicate a problem? Of course it does!" *Assignable cause* and *common cause* variation can now be pointed out. Are the control limits outside of the specification limits? If so, "Class, is there a problem with this product that we have manufactured? Is the process habitually making product out of control, according to the control chart? Is the customer going to be happy with this product? Perhaps or perhaps not!" Are some of the subgroup points outside of the process control limits?

Discuss "out of control" and "in control." A good analogy for explaining in- and out-of-control processes is to use examples of in-control and out-of-control people. Treat the process as a person. This can also help to explain assignable and common cause variation. All people are different, right? This is normal. But, some are just extremely different and may be out of control. A really crazy, out-of-control person is someone that we do not want to associ-ate with, right? Do we want to have out-of-control processes? Of course not! What about the process that has subgroups right up to the control limits? Does anyone in the class want to associate with someone who is just barely in control? Stay away from out-of-control people and processes. If you can, avoid people and processes that are almost out of control. Your stress levels will be much lower!

Participants should be able to at least distinguish the difference between in- and out-of-control conditions, as well as in- and out-of-specification conditions. It is important to be certain the class participants understand the difference between specification and control.

Those subgroups that are outside the control limits when viewing the control chart are usually due to assignable cause. They should also be able to see the common cause variation as the different subgroups are plotted that are inside the control limits. Other concepts to grasp are

- To reduce common cause variation, action is needed on the process as a whole or the system by management and the common cause variation is never entirely eliminated since nothing is perfect.

- To eliminate assignable cause variation, action is taken on specific issues within the process.

The root cause of assignable variation may be very simple, such as a data entry error. Remind the class of the six parts of the Cause and Effect or Ishikawa Diagram. The root cause will always be in one of these six parts of the process. The time spent on discussing variation and causes is dependent on how much time is available to teach the class. However, since variation is the root cause of all quality issues, the discussion on variation is usually well worth it.

Relate the control chart back to the normal bell curve. Approximate a bell curve on the control chart based on the nine subgroups of data produced by the class. Remember to remind the class that the end points of the curve represent the control limits. It is important that the participants grasp that the control limits are the extent of the process' normal variation, regardless of the graphic representation of the process (control chart or bell curve).

Most students will have difficulty in grasping the difference between control, as in control charts, and specifications. Specifically, control limits and specification limits. A simple means to remember and understand is to think of the specification limits as what the design engineer determined the product tolerance should be. However, design engineers do not live in the real world. Specifications may also be set by a customer. However, in both instances, the specifications or the process tolerance may need to be adjusted to reflect reality. The common cause variation may not allow the process to run within the specification limits. This is just the natural state of the process. Remember, the specifications should be considered as theoretical limits. Then think of the control limits as calculated. Control (c) is calculated (c). The control limits represent the real world of the process as based on real measured data.

Can a process that is out of control be brought into specification? Of course it can. But one has to consider the value of this activity. We will discuss process capability, or the ability of a process to meet the specification later in the chapter. However, if the process is capable of meeting the specification, but is not, then sometimes it is a simple adjustment of the process. Manufacturing processes are made up of machines (assembly is not considered to be real manufacturing). A machine or machines can be slightly adjusted to bring a capable process into control. However, it is important to note that adjusting a machine is adding variance to the process. The state of control of the process will get worse after adjusting, before it improves. Remember that variation is the cause of all quality problems in a process. When adjusting, be very patient and make the adjustment very slowly.

Then, what if a process is not capable of meeting specification? If there are multiple assignable causes, at least one will be the lack of operator training. Training is one of the biggest causes of problems in the manufacturing world and it crosses cultural boundaries, but that is a topic for another book. If not addressed at the time of the cause, these multiple assignable causes will typically appear over time as common cause variation. For example, an assignable cause could be data entry error, computer glitch, or machine jamming. These assignable causes should be corrected right away and the data that was entered into the SPC software should be edited or removed. Some assignable causes take much more work and involve more time, such as a machine break down. The machine has to be fixed or replaced and this takes time, as well as possibly being expensive. The loss of production can usually be considered as expensive. The action to take is a determination of risk versus the value.

Think of common cause variation as the "noise" of the process. It is just the way the process is. To eliminate common cause variation, it will take action on the process system. This is normally time consuming and possibly expensive (a subjective term). Again, one has to consider the value gained by reducing the common cause variation. Often it just has to be done. For example, a machine with excessive wear, but still functional, has to be rebuilt or replaced. A major operator training program has to be instituted. Components or materials used in the process have to be changed.

At this point, it is good to summarize what has been presented so far. Keep hammering away at the concept of variation from perfection. Relate the discussion back to the center point of the process, which is the process average, or grand average. Remind the class that as one moves away from the process average, the variance is increasing. The measurement of variance is the standard deviation from the process average, which is measured in sigma units. The greater the value of the standard deviation or sigma is, then the greater the process variance. Draw some bell curves that represent a high value and a low value for the standard deviation (Figures 3.22 and 3.23). Again, discuss that there are three units of standard deviation or sigma above and three units below the process average. The process has a total of six standard deviation or sigma units of variation.

Now it is time to calculate the standard deviation for the data generated by the class. Since the control limits are already known and since the control limits are the limit of the process, then there are six units between the control limits. Simply determine the difference between the control limits by subtracting low from high and divide by six. Avoid the statistical calculation for the standard deviation. Keep this portion of the presentation simple. Remember, this is a lot of material to absorb.

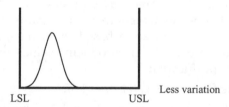

Less variation

LSL   USL

FIGURE 3.22

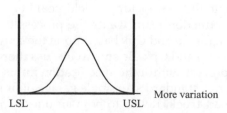

More variation

LSL   USL

FIGURE 3.23

Part 8: Process Capability

"Class, is this enough information to monitor the process?" I hope that at least the class has been able to grasp the value of monitoring the process by using SPC. The class needs to grasp that the desired state is a process that is in control and within specification. That is, with the control limits within the specification limits on the control chart.

Is it enough information to describe to another party the state of the process over the telephone? How do the control limits relate to the specification limits? Can one describe this relationship, in simple terms, over the telephone? Some participants will try to do it. Allow the class to keep trying, but steer the class to the difficulty of describing a graph or chart over the telephone. "Class, wouldn't it be nice to be able to describe the state of the process numerically, instead of having to rely on a picture?" Discuss this concept of in some more detail at this point. Remember my discussion with my friend Mike in Chapter 2.

Define *process capability* and the *process capability indices Cp* and *Cpk*. A generality that I use to describe the difference is to define Cp as more theoretical and Cpk as more of a representation of reality. Be sure to include the key numerical points of 1.0, 1.33, and 1.67.

- A Cp or Cpk of 1.0 means the process is just able to meet specification, but there is absolutely no margin for error.
- A Cp or Cpk of less than 1.0 means the process cannot meet specification 100% of the time. Remember that SPC is sampling. Even if all the samples are within specification, based on the data, with an index of less than 1.0, the process is still producing some unmeasured items that are out of specification.
- A desired state for Cp is greater than 1.67.
- A desired state for Cpk is greater than 1.33. The difference is that Cpk takes into account how well the process is centered between the specification limits.

Draw some bell curves, with specification limits as goal posts. Again, use the analogy of a football team kicking a field goal between the goal posts. Discuss how the production team wants the process to score a field goal. Approximate a value for Cp and Cpk based upon the drawings. Use these to distinguish between Cp and Cpk. Be sure to discuss thoroughly the concept of centering of the process within the specification limits. I also use an analogy of a garage door. A standard size single garage door is not large enough to fit an over-the-road truck. Relate to how an out-of-specification process will not fit within the specification limits, just like a big truck will not fit through a standard garage door. Now, with the same garage door, discuss whether it is easier to drive a 1972 Cadillac Eldorado or a Volkswagen Beetle into the garage (based only on size!). Therefore, it is easier to fit a process with a narrow width or narrow control limits or low process variance, into the specification, rather than a process with a large process variance. Draw a final bell curve, with specification goal posts. Use the data from the class to draw the process bell curve. Discuss whether Cp and Cpk will be greater than or less than 1.0. If time allows, calculate the Cp and Cpk for the data generated in the class.

Process Capability

- Helps a team answer the question, "Is the process capable of producing within specification?"
- Helps to determine if there has been a change to the process.
- Helps to estimate the percent of product or service not meeting customer requirements.
- Cannot necessarily tell you if your process is in control or not.
- Will quickly describe if a process is meeting specification.

Part 9: Calculating Cp and Cpk

Calculating Cp (Simple Process Capability)

1. Obtain the upper specification limit (USL) and the lower specification limit (LSL) or have the process tolerance.
2. Obtain the control limits from the Xbar chart.
3. Calculate the process capability:

$$Cp = \text{Specification width} / \text{Process width}$$

or

$$Cp = (USL - LSL) / 6 \text{ sigma}$$

or

$$Cp = (USL - LSL) / (UCL - LCL)$$

Cp relates the spread of the process to the specification width. It does not look at how well the process is centered within the specifications. This concept of centering is important as we will show as we calculate Cpk. Remind the class of some of the bell curves that were between the goal posts, but off center.

Calculating Cpk (Real Process Capability)

Cpk is a measure of the process variation with respect to the specification tolerances and the location of the process average relative to the specification target. Cpk measures the *actual* performance of the process in relation to the specification. Note the symbol for sigma is σ.

In mathematic terminology, the formula appears this way:

$$Cpk = min\big[(Xdbar - LSL)/3\sigma, or (USL - Xdbar)/3\sigma\big]$$

Fortunately, there is a step by step process for calculating Cpk that explains the terminology:

1. Obtain the process average (Xdbar), and the process range average (Rbar) from the Xbar and R charts or in this class from the data matrix.
2. Obtain the USL and the LSL.
3. Calculate the process standard deviation, sigma or σ
   i.   sigma = UCL – LCL / 6
   or
   ii. sigma = Rbar / d_2

   d_2 is another mathematical constant as was A_2. For a subgroup of five, $d_2 = 2.326$. For a subgroup of six, $d_2 = 2.534$.
4. Calculate the process capability:
   i.   Cpk(Xbar – LSL)/3σ
   ii. Cpk(USL – Xbar)/3σ
5. Choose the lower value of the two results. This is the Cpk or true process capability index.

Part 10: Xbar and R Charts

I have focused on the Xbar charts because these are the most difficult to understand, provide the most information and if doing manual calculations, the most difficult to construct. However, with SPC software and when doing

Xbar chart

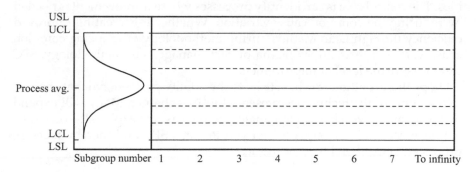

FIGURE 3.24

Range chart

FIGURE 3.25

manual charts, there is always a range chart. This is a simple run chart that plots the subgroup range directly in line with the subgroup averages. In the software, the Xbar chart is on top and the range chart is directly below as in Figures 3.24 and 3.25.

Part 11: Summary

The class has learned how to use sampling and data to monitor a process and to evaluate product quality. It is a good idea to discuss the value of using the SPC information. Mass inspection or 100% inspection is very expensive. In addition, it is not that accurate. SPC will provide a good picture of the process. The long-term advantages of using SPC are many. Some are: a more knowledgeable workforce, fewer defects, less rework, less overtime, higher profits, and greater customer satisfaction.

SPC is an excellent tool to use with Lean Manufacturing or Six Sigma activities. The control charts can identify processes where improvement is needed by reducing the common cause variation. Whether it is creating improved efficiency through Lean Manufacturing methodology or reducing common cause variation through Six Sigma project management methodology, SPC can measure the level of improvement.

I hope these methods are beneficial to the quality professional that chooses to use some of the techniques presented in this chapter. Many will expand on the ideas and will have better analogies and cases to explain the concepts. If I have been able to help a few of you to teach SPC and thereby increase product quality, then I have been successful.

4

Variation in the Real World

All of life is education and everybody is a teacher and everybody is forever a pupil.

Abraham Maslow

One of the most important concepts that I want to get across is that when a process is changed, it will get worse before it will get better. This is regardless of how justified the reason for the change. It does not matter if the process is in control or out of control. Why? As discussed in earlier chapters, it is adding variance to the process. Making a change or adding a variable or more than one variable causes problems in the process. An in-control process can go out of control when a change is made. Only make changes when there is *problem variation* which will be described in this chapter.

Remember, the level of statistical control is the natural state of the process. Therefore, whether a process is in or out of control, whether good or bad in relation to specifications, it is the state of the process. However, one does not usually want their process to be out of control for any more than a very brief period of time. If the process is out of control, this means there is excessive variation and there is an assignable cause or causes. Action should be taken to bring the process into statistical control as shown on the control charts. That is, all subgroup data must be in between the control limits. If not, a state of statistical control does not exist; the process is considered to be out of control. It may be slight or for a short duration of time, but it is still out of control. However, one must remember that any action taken to bring a process back into control will initially have even greater variation. That is, it will get worse before it gets better. Consider the value of taking or not taking action when a process is minimally out of control or out of control for only a brief period of time.

To bring the process back into control, action needs to be taken on the assignable cause. However, sometimes there may just be a slight blip in the process and the process will correct itself. In this instance, no action needs to be taken. Other times, there is a process that is perpetually out of control on the upper side, the lower side, or, the worst case, is out of control on both the upper and lower sides. This is problem variation and action needs to be taken.

Imagine a high-speed soft drink bottling line. There is a machine to load the bottles onto the line, a revolving filler machine with multiple filling stations, a capper, a machine to apply the date and lot code, a carton packer, a case packer, and a palletizer to stack the cases. Then, there are many conveyors to connect all of these machines. Synchronization of all these parts to the process is very important for maximum efficiency. If a line operator speeds up the filling machine and conveyor coming out of the filler, the line may jam at the capper. Then, this high-speed line stops. There is a mess to clean off of the line and any partially filled or uncapped bottles must be removed. Why did this happen? The operator added variable to the line by speeding up the filling machine and thus put the line out of synchronization. Quality was also affected in addition to the line efficiency. Consider how many bottles had sticky soda on the outside after the line jammed. In this example, the assignable cause is operator adjustment or an untrained operator. That is not a difficult cause to rectify, but the training, especially if it is this type of SPC training, can be time consuming.

Sometimes, the action to bring the process into control will be expensive. Examples are new equipment, equipment repair or refurbishing, extensive personnel training, process restructuring, automation, and material upgrades are just a few examples. Other times, the action may be minor and therefore less expensive. Perhaps a machine just needs minor adjustment, minor repair, or routine preventive maintenance that is already scheduled. An operator may just need a brief refresher of their SPC training.

Regardless, once it is certain that a process is out of control, action needs to be taken. I call this condition problem variation. There is always an assignable cause when a process is out of control and causing problem variation that needs to be reduced or eliminated. Work on one cause at a time, then let the process run to observe if there is any improvement as shown on the control chart. Then work on the next cause and again let the process run for a while to observe the change to the level of control. Repeat this type of activity until the process is in control which is the desired state where all the variation is common cause. Be patient ... take it slowly.

If one has a process that is in statistical control it is usually best to just let the process run as it is. There are some exceptions to this. For example, if a customer requires strict adherence to their specification limits, and the control limits as shown on the control chart do not meet the customer specification limits, then action needs to be taken even though the process is in control to bring the process into the customer's specification range. Remember the definitions of quality in Chapter 3. Again, as action is taken to bring an in-control process into specification, due to the changes it will likely initially have an increase in process variance. Just let the process run and it will settle into the new parameters. Another example of when to take action on an in-control process is if there are legal issues involved. Consider a consumer product that has a label claim of 20 ounces. If the process is consistently filling the container at less than 20 ounces, even though it is in control, then action needs to be taken to meet the label claim of 20 ounces. Note there is no legal issue to fill a container to a level higher than the label claim. This is just giving away free product, which is not illegal. If the process is consistently filling higher than 20 ounces, one has to consider the cost of bringing an in-control process to within the upper specification limit. It is usually not a matter of simple adjustment or this action would have already been taken. Rather it involves taking action on common cause variation since the process is in control. This can be rather expensive. The cost or risk versus the value gained has to be evaluated to determine if the project is worthwhile.

Sometimes, if the out-of-control condition is slight and for a short duration, one can continue to let the process run as it is without making changes. I know this will be considered as blasphemy to other professionals. There may have just been a slight blip somewhere and the process, if allowed to run, will correct itself. I will continue to preach to avoid making adjustments or changes to the process. Be patient! Just remember that adjusting is adding variance which can cause the process to go out of control. Adjust or make changes only when it is determined to be necessary. Only adjust or make changes when there is problem variation.

Let's consider our control chart. I intentionally left out the six units of standard deviation or sigma in Chapter 3. These are normally part of a control chart as shown in Figure 4.1. The units of standard deviation from the process average are shown by the dotted lines and the upper and lower control limit lines. Remember the UCL = the process average plus three units of standard deviation or sigma units and the LCL = the process average minus three units of standard deviation or sigma units.

Starting from the process average and moving up, the first dotted line is plus one standard deviation units of variance from the process average, the second dotted line is plus two standard deviation units of variance from the process average and the UCL or upper control limit is always plus three units of standard deviation of variance from the process average. Starting from the process average and moving down, the first dotted line is minus one standard deviation units of variance from the process average, the second dotted line is minus two standard deviation units of variance from the process average and the LCL or lower control limit

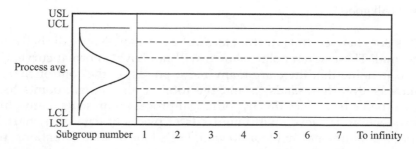

FIGURE 4.1
A standard control chart.

is always minus three units of standard deviation of variance from the process average. As discussed in Chapter 3, notice the units of standard deviation are always of equal width whether on the control chart or the bell curve.

Now let's plot some hypothetical situations. Let's assume each subgroup data point is from our high-speed soft drink production line and is a sample of five pieces of product measured once every hour. Therefore, each plotted subgroup represents one hour of production time. Also, note that the line does not stop when the shifts change. Therefore, all eight points represent a specific shift of eight hours of production. In other words, samples are taken every hour, measured, and the data entered into the SPC database, either manually or automatically. Assume a 24/7 operation where the production lines rarely stop.

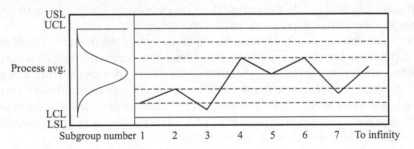

FIGURE 4.2
(March 6)

What do we see in Figure 4.2?

- All points are inside the control limits and inside the specification limits.
- Total variance is a little over three standard deviation units.
- Is all good?

This does happen and is fortunately quite common. Notice the bell curve is normal. These seven points shown will not have a normal curve. One has to imagine that there are many more points on the control chart ... hundreds, even thousands ... and the total of all the data points has a normal bell curve, since this hypothetical process represents many shifts of production. However, with hundreds of pieces of data on a chart, all sections or shifts may not look so well. Let's look at some situations, that is, points in time from the same very long control chart for the same production line.

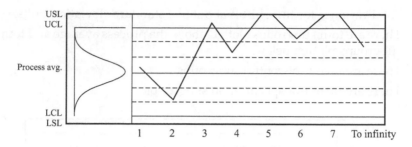

FIGURE 4.3
Control chart indicating shift in production (March 19).

What does the shift of production as indicated on this section of the control chart tell us as shown in Figure 4.3?

- Seven out of eight points are above the process average.
- Two points are outside the upper specification limit.
- Three points are out of control since these are outside of the upper control limit.
- The process is running on the high side.
- There is general trend to the upper side of the control chart.
- Overall, process variance is over six units of standard deviation, which is quite a bit of variance.

What action should a production operator take with the control chart data for their shift as shown in Figure 4.3? Before we answer, let's look at some other situations.

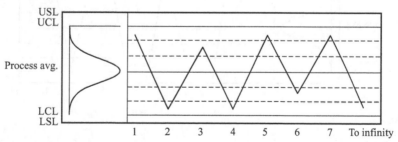

FIGURE 4.4
(May 20)

What does Figure 4.4 tell us? Should action have been taken during this shift? Are any of these issues significant?

- First of all, one should see that there is a lot more variance on the chart for this shift than shown in Figure 4.2 and less that the total amount as shown in Figure 4.3.
- The variance is approximately centered around the process average.

- Total variance for this shift is about five standard deviation units.
- There are four subgroups plotted above the process average and four subgroups plotted below.
- There are no out-of-control subgroups.
- There are no out-of-specification subgroups.

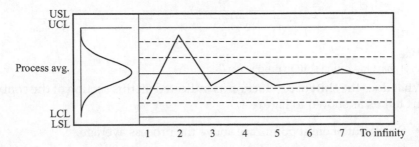

FIGURE 4.5
(June 12)

What does Figure 4.5 tell us? I hope one can see that no action should be taken. Everything is fine! We'll discuss this more later in this chapter.

- The shift started out with more variance and then settled down with minimal variance for the last six hours.

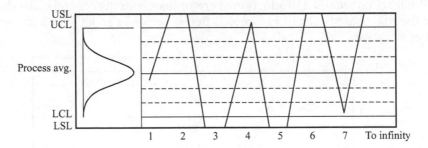

FIGURE 4.6
(June 12)

Let's assume that Figure 4.6 represents the shift immediately following the shift that is represented by Figure 4.5. What does Figure 4.6 tell us?

- One should see a completely out-of-control situation in this shift.
- Variance is excessive!
- Six out of eight subgroup data points are out of control.
- Five are off the chart, and they are out of specification, both low and high.

Clearly, there is a problem! Why did the process go out of control on the back-to-back shifts as shown in Figure 4.6 when it was fine in Figure 4.5?

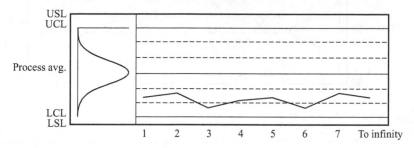

FIGURE 4.7
(July 17)

What does Figure 4.7 tell us?

- Variation is minimal.
- All subgroups are on the lower half of the control chart.
- All subgroups are below the minus one unit of standard deviation line.

Is this significant? Should any action be taken?

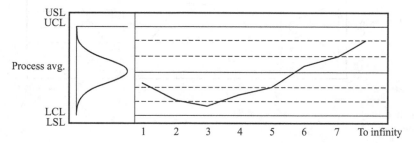

FIGURE 4.8
(July 20)

What does Figure 4.8 tell us?

- There is a trend in a positive direction toward the upper control limit. Where will subgroups nine, ten, and eleven be?
- Variance, if measured between subgroup number three and number eight is slightly over four units of standard deviation of variance, which is quite a bit, but not extreme. But, variance from one subgroup to the next is not excessive.

Is any of this significant? Should any action be taken?

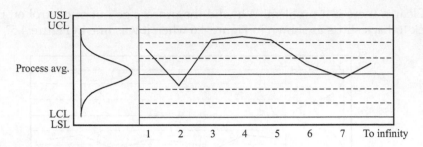

FIGURE 4.9
(August 10)

What does Figure 4.9 tell us?

- Six out of eight subgroups are in the upper half of the control chart.
- Total variance equals about three units of standard deviation.
- Subgroups three, four, and five are more than two standard deviations from the mean.

Is any of this significant? Should any action be taken?

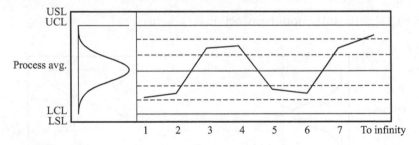

FIGURE 4.10
(September 20)

What does Figure 4.10 tell us?

- Total variance is about four standard deviation units.
- All eight subgroups are more than one standard deviation from the process average.
- Subgroups six, seven, and eight are showing a short trend toward the upper control limit.
- Is there a pattern?

Is there anything that is significant? Should any action be taken?

Before answering the question as to whether action should have been or should be taken based on the data shown on the control charts in

Figures 4.2 through 4.10, let's consider some statistical textbook type of information.

In the *Journal of Quality Technology* (October 1984), Lloyd Nelson* published a set of rules for process control to determine out-of-control signals known as the Nelson Rules. These rules apply to the Xbar or variable values chart.

1. One point is more than three standard deviation units from the mean.

2. Nine (or more) points in a row are on the same side of the mean.

3. Six (or more) points in a row are continually increasing or decreasing.

4. Fourteen (or more) points in a row alternate in a direction, increasing then decreasing.

5. Two (or three) out of three points in a row are more than two standard deviations from the mean in the same direction.

6. Four (or five) out of five points in a row are more than one standard deviation from the mean (process average) in the same direction.

7. Fifteen points in a row are within one standard deviation of the mean (process average) on either side of the mean (process average).

8. Eight points in a row exist, but none within one standard deviation of the mean (process average), and the points are in both directions from the mean (process average).

The American Society for Quality (ASQ) has also identified some out-of-control signals.

1. A single point outside the control limits.

2. Two out of three successive points are on the same side of the centerline (process average) and farther than two standard deviations from it.

3. Four out of five successive points are on the same side of the centerline (process average) and farther than one standard deviation from it.

4. A run of eight in a row are on the same side of the centerline (process average). Or 10 out of 11, 12 out of 14, or 16 out of 20.

5. Obvious consistent or persistent patterns that suggest something unusual about your data and your process.

---

* Dr. Lloyd S. Nelson (1922–2013): An American statistician, he expanded on Walter Shewhart's principle and published, in the October 1984 *Journal of Quality Technology*, the Nelson Rules for out-of-control conditions as seen on a control chart.

Well, let's see if any of these rules apply to Figures 4.2 through 4.10 as discussed below.

- Figure 4.2: Borderline Nelson #5 and ASQ #2 (point number one is exactly on the minus two standard deviation line)
- Figure 4.3: Nelson #1, Nelson #5, Nelson #6, ASQ #1, ASQ #2, ASQ #3
- Figure 4.4: Nelson #5, Nelson #8, ASQ #2, possibly ASQ #5
- Figure 4.5: No issues
- Figure 4.6: Nelson #1, Nelson #5, Nelson #8, ASQ #1, ASQ #2, possibly ASQ #5
- Figure 4.7: Essentially Nelson #2, Nelson #6, Nelson #8, ASQ #3, ASQ #4
- Figure 4.8: Nelson #3, ASQ #5
- Figure 4.9: Nelson #5, ASQ #2
- Figure 4.10: Nelson #8, possibly ASQ #5

I hope all can agree that for the shift represented in Figure 4.5, there are no issues and no action to take. Then I hope all can agree that for the shift represented in Figure 4.6, something is wrong and some type of action is needed. However, for all the rules in all the statistics text books and the rules shown here, I say … hogwash! If the process is in control and in specification, leave it alone! This is the real world. Go back to the very first paragraph of this chapter. Only take action when there is a problem!

Let's take the situation going from the shift represented in Figure 4.5 to the shift represented in Figure 4.6. What happened? Something changed. There is a problem. What are the possibilities? Well let's start with the Ishikawa Fishbone terminology: environment, materials, methods, machines, manpower (people), and measurement.

- Environment: Highly unlikely that the environment changed enough to cause such a dramatic and immediate effect on the process at the exact time of the shift change.
- Materials: Possibly, but it seems pretty coincidental that the material change happened right at the start of the next shift.
- Methods: High-speed production lines are not changed. The overall method of filling and packaging the soft drink bottle does not change.
- Machines: Possibly, but machines usually do not cause such up and down variance. In Figure 4.6, we see the process going out of control high, then low, then high, then low, etc. If a machine was not functioning properly then the line would have been shut down for maintenance, rather than running an entire shift out of control and out of specification.

- People: Hmm. Since there was a shift change, then people should be immediately considered. The production as shown in Figure 4.6 is a classic picture of operator adjustment. The process was adjusted ... perhaps it was the line speed to get out more cases and look good to the boss. Oops! The process is out of control on the upper side. Adjust the line speed again. Oops! Now the process is out of control on the low side. Adjust the line speed again. Oops! The process is out of control on the upper side again. And so on through the entire shift. Had the operator left everything alone, we probably would see a chart similar to Figure 4.5.

- Measurement: Possibly, but unlikely that measurements would be taken incorrectly for an entire shift. If the operator was uncertain, they would normally ask for help ... even unofficial help from a friendly coworker.

What I'm trying to demonstrate graphically is what I have preached for a couple of decades: if it ain't broke, don't fix it. If the process is in statistical control and in specification, leave it alone. Often there are situations where even if the process is running out of specification, but is in statistical control, leave it alone. Only make changes when there is problem variation such as in Figure 4.6. That change would not be due to the process. Rather it is training, remedial training, or replacement of the line operator. The process is fine as shown in Figure 4.5 and as represented by a normal curve with hundreds of shifts of production.

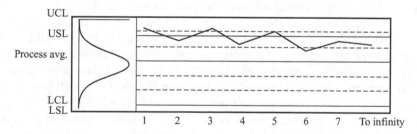

FIGURE 4.11
(March 17)

In Figure 4.11, let's again consider our high-speed soft drink production and packaging line. Consider what is measured is the amount of soda in ounces. With the information just presented and the discussion on Figures 4.2 through 4.10, what does this control chart of production for a shift on March 17 tell us?

- The process is running in the upper part of the control chart.
- The process is in a state of statistical control.
- Currently some bottles are being filled with a little extra soda. So what!

- Absolutely no action should be taken! Let the line run and produce a shift of high-quality soda.
- There is no problem variation!
- If an adjustment is made or the line is stopped, then the result could likely look similar to Figure 4.6 or possibly Figure 4.3.

To further discuss problem variation, I like to compare two types of variation; problem variation and common variation, when considering situations graphically shown on a control chart in the real world. For common variation, I use the classic definition as in Chapter 9. It is just the normal state of the process when in statistical control. It is the background "noise" of the process. Problem variation is when one has to take action. Figure 4.6 has problem variation. Some kind of action clearly needs to be taken and right away. In Figure 4.3, there is problem variation, but I would evaluate and take extra samples before taking immediate action. It could be that the process will sort out whatever the issue is. I would not take any action in the other illustrations Figures 4.2, 4.4, and 4.7 through 4.11. I know statisticians will have a problem with my beliefs. But, fortunately, I'm not a statistician, although I understand statistics. My beliefs are entirely focused on events experienced in the real world of manufacturing.

If I was a manager or in a leadership role, I would have stopped the line early in the shift, as shown in Figure 4.6. My action would be to find out what happened and what changed. I would tell the operators to relax and discuss the recent events. I'd bring the process back to where it was when the shift started and start the line again. It would likely start out similarly to what is shown in Figure 4.6, but if allowed to run a while, it should return to the state similar to that of the previous shift. This is considering that the assignable cause was operator adjustment. Also, there is the subject of scrap. Rather than running a whole shift as shown in Figure 4.6, which produced a lot of defective soda, by stopping the line I would have reduced the amount of scrap and defective soda produced on the shift. There would still be some scrap and defects to discard. But, the amount from 30 to 60 minutes is a whole lot less than that of the entire shift. In addition to the defects and scrap product produced on this shift, it is highly likely the shift efficiency was low and the case output was significantly lower than average. By sorting things out and getting the process back into control early in the shift, I would have had good line efficiency and quality for the remainder of the shift.

Now let's discuss common cause variation. Looking at the bell chart for all of the data entered into the SPC database for this process, one can clearly see the process is operating within the specification limits. One can drive the car through the garage opening without much difficulty. Cp is definitely greater than one. Cpk is also greater than 1, but may not be equal to or greater than 1.3, although it will be close. Remember that Cpk takes into account the centering of the process in relation to the specification limits. This process is running closer to the lower specification limit than the upper. If I was assigned the task of raising the Cpk value, which would involve reducing the common cause variation and bringing the control limits closer together, I would evaluate the better portions of the SPC control chart.

Looking at Figures 4.5, 4.7, 4.11, and possibly Figure 4.9, minimal variation is seen. I would evaluate these shifts to determine why they have such minimal variance, that is, the width between the high and low measures or range for those shifts. Are they the same scheduled shift ... first shift, second shift, or third shift? Perhaps the data is from a weekend shift. Did these shifts operate right after some type of scheduled maintenance? Are there different suppliers of materials, perhaps the bottles? Assuming the line is used to fill different flavors of soda, is there a flavor that tends to allow the line run better, which would primarily be the filling machine? Is there an environmental factor? What was the temperature and humidity on that shift? It is very easy to obtain this data for the external ambient temperature, humidity, and the dew point from the National Weather Service. For the interior of the building, it can be more difficult or impossible to get the temperature and humidity data. However, the exterior environmental conditions influence those of the interior.

Once I was able to determine the attribute(s) that allowed the line run with the lesser amount of variance, I would then determine what it would take to apply this information to the other shifts. It sounds simple, but it can take a lot of study since there are so many potential variables in a high-speed liquid filling line for soft drinks that we are using as an example. I once spent significant time over a three-month period evaluating a similar type of production line to determine why there was such a wide range of process variance. Sometimes, the line would run great. Other times, it was sporadic. What was the cause? In this instance, the bottles were plastic and there were three different bottle suppliers with a total of five different bottle molds that produced the bottles. Once I was able to determine which bottle ran the best, I was able to obtain the approval for a capital expenditure to renovate or replace the other four molds. Was it worth it? Well, prior to my being able to obtain bottles that were all the same, the average case output on this production line was about 1,700 cases on an eight-hour shift. Afterwards, it was common to for the line to produce over 5,000 cases in the same eight-hour shift. The third shift was no longer needed to run this production line and they were assigned another line. The second shift only ran this line on occasion. They usually ran about four lines on the second shift, so the resources were also usually moved to another production line. This line always ran on the first shift and guess what? The quality of the end product also greatly improved with the line efficiency. Production line efficiency and quality are directly proportional. It was definitely worth the capital expenditure, and there was a measurable cost savings or return on investment.

What would the next step be once the process in our example was modified and ultimately had less total process variance and a higher Cpk and Cp?

Well, one could go to another process in need of improvement, but what if senior management wanted to further reduce the total process variance on this same production line? It would be possible, but at some point the law of diminishing returns will kick in. It is not likely to get a reduction in process variance as great as the first time, but it will at least be the same amount of work and probably more, as well as the same amount of cost and probably more. Perhaps the company wants to invest several hundred thousand dollars in a new, larger, and faster filling machine. What will be the value of this investment? Remember the law of diminishing returns is a fact at some point. The value obtained must always be evaluated against the cost or the risk.

For this chapter I will leave you with this:

> Listen, listen, listen to the people that do the work!
>
> **H. Ross Perot**

This is something many managers in the real world are not very good at.

5

War Stories

There's only one corner of the universe you can be certain of improving, and that's your own self.

Aldous Huxley

Experience: that most brutal of teachers. But you learn, my God do you learn.

C. S. Lewis

All right, I admit it. I've got a lot of experience, which makes me kind of an old guy. About 40 years' worth of experience at the time of this writing. Another quote I know, that I cannot credit to the right person is that, "Youth is given ... age is achieved." Well the same could be said about experience: it is achieved. There have been areas in this book where I've been critical of managers or other so-called leaders. It is because there are just so many bad ones that I have been so fortunate to be associated with. For one period of employment, I worked for a CEO that was a complete sociopath, an opinion shared by all management and leadership staff. However, my time spent with good managers and good engineers was well worth it. While few in number, they mentored and nurtured my career. Without knowing, they molded me into something useful. Although I still managed to make mistakes, I am very proud to have broken out of the ranks of a manager to the director level. My many years of experience are the source of my many stories.

Metal Stamping

Quite a few years ago, I was working with a metal stamping factory. They used big machines, called presses to stamp pieces of flat metal into a specific shape through the use of a die or tooling. The bigger the press, the bigger the part it is able to stamp out. The presses are rated in tons, with the bigger the machine, the greater the tonnage rating. In this situation they were producing large brackets that were about two feet long and six inches wide out of heavy gage, thick steel. Quality control (QC) samples were taken once every hour and the resulting measurements were entered into an SPC database. It took three operators to run the process. The QC technician was independent and sampled product from all the various presses.

The process was in statistical control when I started working with the company and the Cpk for the two dimensions we are going to discuss were about 1.45 for one critical dimension and 1.57 for a noncritical dimension. Over a period of time, it was noticed that the process had gradually moved toward the lower specification limit for these two dimensions. There were several other dimensions that QC checked, but only these two became a concern.

Several months later, the Cpk for the critical dimension had dropped to about 0.98 and the noncritical dimension was about 1.11. The process was still in statistical control for these two measurements as shown on the control chart, but the current Cpk indicated that a few defects were in fact being produced. For this type of operation, the SPC control charts were a means to watch tooling wear, to know when the tooling should be refurbished, as well as monitoring the stamping process for out-of-specification conditions. Tooling wear was always indicated by a gradual trend over a period of a few years toward the upper specification limit. Engineering was quite puzzled over the fact that these two measurements had moved toward the lower specification limit and that there was little change in the Cpk or control charts for the other dimensions. The tooling die was still early in its lifespan. The production manager's response was to keep running, "There's nothing wrong. It's still in control." Production managers generally have one goal and that is to get product out the door and on the way to the customer. Very few really care about product quality. Stopping production is unheard of. "But the customer needs the product" is their battle cry! Yes, but the customer does not want the product with defects! Shipping bad product on time is not the answer.

The manufacturing engineer, while knowing in his gut that something wasn't right, did not stand up the production manager. He had sampled many pieces, but could not find one where either dimension was actually out of specification. The engineer that designed the tooling also agreed there was nothing wrong, because he designed the tooling and he did not make mistakes.

Yet there was something wrong. Guess who determined it? That's right, the customer, the ultimate judge of quality. Remember the critical dimension had a Cpk that had slipped to about 0.98. A few brackets out of every batch were not meeting the specification on this dimension and it was gradually getting worse. The company had to take a lot of brackets back that the customer had accumulated that did not fit their process. In order to keep the customer happy, they had to inspect every new bracket for the critical dimension. This is called sorting and seems to be the ultimate dream of many production managers to solve quality problems. It is very expensive as it is labor intensive. Also, see Dr. Deming's point number three and consider Dr. Taguchi's Loss Function.

So what was the problem? There was a design flaw in the tooling used to stamp out the bracket. This is the same tooling designed by the engineer that never made mistakes. The result:

- The company had to take back a lot of defective brackets.
- More defective brackets were found in their 100% inspection.
- Labor costs dramatically increased till the tooling was reworked.
- And most of all, there was an unhappy customer.

What is the lesson? SPC and control charts do not lie! The data, assuming a correct measurement, is the reality of the process or, in this case, these dimensions. Trust the control charts. Trust the Cpk. Trust what these two indicators are telling you!

Plastic Injection Molding

One time, I was working with a company that had a product in a plastic case that was about six inches square with slightly rounded corners, with a top half and a bottom half that snapped together with the mechanism of the product inside. In the center of the bottom half, a three-inch square piece of double stick foam tape was attached. When the end customer received the product, they would peel off the covering of the double stick tape and attach this product to the surface of choice at their location. These could be attached facing any direction, including upside down.

A common customer complaint was that the product could not stick to a surface because the base had warped into a slightly concave shape. Since the tape was only about one-eighth inches thick, any warping greater than that would cause the product to not stick to the surface it was applied to. If a lot of muscle was used, then the product would crack. There were lots of customer complaints.

The company that manufactured these had gone through Lean Manufacturing. They molded the top and bottom pieces of the plastic housing at their location. The molding process fed into the assembly process. Since the assembly could process faster than the molding process, the molding process ran 24/7. This is not unusual in molding, as shutting down a molding machine and starting it up again is rather time consuming. However, being Lean meant there were times when there was slow demand or sales for a certain color and the molding machine would be temporarily shut down.

Pulling fresh samples of the base piece showed some extremely slight warping, but nothing that was significant. Different colors showed a different degree of warping, but in all colors the warping was very slight and not enough to cause the problem the customers were complaining about. By happenstance, for a reason that I forget (remember, a career of many years), we went back a few days later and re-measured the samples that had been taken of the base. The warping had increased and the increase was significant. It was enough to prove the validity of the customer complaints.

Why was this happening? One has to understand molding, but one can imagine there are a lot of variables involved in plastic injection molding. First, there is the mold itself, which will wear after millions of shots of melted plastic resin. Then, there is wear to the actual molding machine. There is the speed or cycle time of the molding machine. That is the time from point A in the injection molding process and back to again to point A for the next injection. Then, there is temperature. The temperature the machine is set at depends on the type of plastic resin, the cycle time of the machine, and the size of the product being molded. It is very important.

Since we now had warped bases after a period of time, we were able to focus on temperature because the molded product samples clearly had continued to cool over time. When a few hundred bases were dumped into a box at the end of the molding machine, there was not much venting and the heat of the molding process was somewhat retained. It was not until after assembly that the individual bases were able to cool enough to warp. But, what had happened to the process? Well, in order to turn up the cycle time to a slightly faster rate, the temperature had to be increased to keep up with this increased cycle time. This was to produce more bases in order to stay ahead of the assembly process or to even reach a point of shutting down and doing something else or going fishing.

SPC was already used in the molding process with samples measured every half hour and the data entered into the SPC database. The molding SPC data was very good, but then they were measuring freshly molded bases. So what was the answer? SPC! Samples of the bases were taken and were marked with the time and date of production. These were held in an open area to cool for 24 hours. The amount of warp was then measured and entered into the SPC database. Molding had to stop their shenanigans and slow their machine down to the specified cycle time and reduce the temperature accordingly. Again, the control chart and the Cpk do not lie. The control chart and Cpk would identify any tendency toward warping, indicating an adjustment in the molding process.

Filling Bottles of Lubricating Oil

At another point in my years of experience, I was working with a company that was filling rigid, small plastic bottles with lubricating oil. These bottles were sealed with plastic screw-on caps that had a liner for sealing inside the cap.

The process had a machine that loaded the bottles that were dumped into a hopper onto a conveyor with the proper orientation. The bottles then went into an eight-stage rotary filling machine. If one has not seen one of these machines, it may be hard to visualize, but I shall try. An eight-stage rotary filler is essentially filling eight bottles at once. As a bottle is picked up by the filler from the conveyor, it starts to fill with oil. In this instance, as the bottle revolved around the filler, it was not completely filled until the final stage when it was put back onto the line conveyor. So, in about three feet of line space, the filler was constantly filling bottles with oil, eight at a time. This is because the filling process must be slow to avoid pumping the oil into the bottle too quickly and getting droplets of oil on the outside of the bottle. The primary objective was line production speed, but oil on the outside of the bottle also would not allow for adequate label adherence.

As the bottles moved down the conveyor, the next step was the capper. The capper also oriented caps in the proper direction from a bulk hopper. The caps came down a chute and as a bottle passed under the chute, a cap was picked up on top of the bottle. The cap was then torqued onto the bottle to the proper torque specification. Cap torque was important because too loose of a cap would leak and too tight of a cap would be difficult to open.

The SPC torque data had gradually gone out of control over time and exhibited problem variation. Some caps were way over-tightened. Maintenance worked on the capper constantly usually not finding any caps that were over-tightened. But, then the operator would get one during their routine sampling and the SPC Xbar chart showed that the process had some problems. Finally, maintenance gave up, as they did not know what else to do.

So senior management budgeted for and ultimately purchased a new capping machine. After the installation, the process was not improved. About this time, I was brought into the issue to form a cross-functional problem-solving team. Why did the problem still exist when there was a new capper?

Our first action was to run the process without stopping ... sound familiar? We were hoping that the starting and stopping with a focus on the functionality of the capper was causing the excessive variance by introducing changes to the process. After all, the process was no worse off than before. The factory was shipping the same level of quality as they had with the old capper. After two weeks, there was essentially no change.

The problem-solving team then went into further brainstorming mode. The administrative person on the team, who happened to be a lady, suggested the problem was with the filler. Maintenance and operations immediately shot this comment down. "No," they said, "the problem is the capper," clearly implying this person did not know what they were talking about. Of course, this person made no further verbal contributions to the project team.

- The capper is not right for the process.
- Run the line faster.
- Run the line slower.
- Keep adjusting the capper.
- There's something wrong with the cap.

The cap supplier and the capping machine manufacturer's representative were both brought in to help. However, knowing something about brainstorming and problem-solving teams, I was intrigued by the suggestion that the problem was with the filler. I started to focus on the filler and did a lot of sampling and observing. What I discovered was that one of the eight filling stations on the filler was leaving a minuscule drop of oil on the thread of almost every bottle that went through that filler head. This little drop of oil caused the threads to be lubricated as the cap was torqued into place. The oil was nearly clear and the drop was small, so it was very hard to see this drop of oil at the speed the line was running. Again, this process was filling lubricating oil. The lubricating oil was lubricating the threads on the bottle so that when the correct amount of torque was applied to tighten the cap, it actually over-torqued because the activity was easier. The cap just slipped over the threads into an over-tightened condition.

There are multiple lessons in this example:

1. Trust the SPC charts! The data does not lie.
2. Never, ever ridicule an idea in a brainstorming session.

3. Don't be too quick to spend money, especially as a capital expense to solve a problem.

4. The most ridiculous idea, from the most unqualified person, may just have some validity.

Manual Powder Filling

Going way back in time, I was associated with a manual powder filling process. The powdered product was sold to the scientific and academic communities. Bottle sizes ranged from 50 to 250 grams. An empty bottle was put on a scale and the scale was tared to zero. This tare was used throughout the filling of that size bottle, regardless of lot size. A second person then capped and labeled the bottles. They then packed the bottles into cases. The unit of sale could range from a single bottle to a case.

The production manager was trying to help justify the cost of an automatic powder filler and asked me to do a study. Of course, SPC was one of the tools I used to evaluate the accuracy of the filling process. Through observation, I noticed that the person doing the filling never even looked at the scale. Rather he/she just approximated the fill level in the bottle based on experience from the first few bottles filled in the batch. Different products had some particle size differences, so the visual fill level could vary slightly between products in the same size bottle.

I have to admit that the filler would develop something of a cadence and was actually pretty fast and accurate at filling within the allowable tolerance. Thinking that my presence was causing a bias, I started sampling at random times, known only to me. I was still impressed with the speed and accuracy of the person doing the filling. However, the more data I plotted, the more my Xbar chart started to show me there were some issues. I should have guessed these, but I was young and naive.

- As break or lunchtime approached, the accuracy declined as the filler was watching the clock more than what he/she was doing.
- First thing in the morning, especially on Monday or the first day of the week, it took some time to get into the rhythm of things as far as accuracy and speed. I actually suggested a paid coffee break with company supplied coffee, but of course this idea was quickly squashed as total blasphemy.
- Then toward the end of the day, and especially on Friday, there was again a lapse in filling accuracy because there was more clock watching.

I heard anecdotally that they finally got their automatic filler a couple of years later. Again, I was truly impressed with the overall accuracy and speed of the people doing the filling. It was a job that I would not have been very successful at. What I want the reader to take away from this story is that:

- Again, the SPC charts do not lie.
- Sometimes it takes a lot of data to get a good, true picture of the process on a control chart.
- People try to do a good job.
- People are human. They are not machines.

Fishing Reel Drag

More recently, I worked with a fishing reel manufacturer. These were high-end, high-quality fishing reels. The CEO told me he wanted consistency. He never once used the word "quality." He stated that when a guy tried the sample reel at the sporting goods store and bought one, he wanted the reel the guy took home to feel the same. That is, he wanted consistency.

I instituted multiple changes at this small factory. All supported the cause of improved quality and efficiency. I knew the CEO was talking about quality when he spoke of consistency. Efficiency was important because a CEO thinks in terms of dollars and as I made changes I could not add bureaucratic waste to the process. Instead, I helped improve efficiency.

Early in my tenure at this factory, I asked the plant manager for permission to go talk to the line workers. "What for?" he asked. "I've already told you how we assemble the reels. And so has the engineer," he added. My response to him was, "Well, guess what? You and the engineer don't assemble the reels in the same manner. And I guarantee you that the line workers have developed their own methodology." Again, much of my success as a consultant was due to the fact that I would listen to the people doing the actual work. I really would listen with rapt attention as I knew they understood where the problems were. I know that I've mentioned this earlier in the book, but it is a fact. In addition, production workers are often starved for someone to listen to their ideas. Often, they think that management just doesn't care. Too often, they are correct. Of course, by working with the line workers I gained a much greater understanding of the assembly process and was able to start making improvements, of which I made several. While my focus had to remain on improving consistency, I was working up to it gradually. I was also able to make a lot of improvements in a short period of time.

I'll share one quick non-SPC story from this factory experience. Debbie had a complete tool box with all the tools needed to do the job. Her tools were either wrapped with fluorescent orange tape or sprayed with fluorescent orange paint. She could spot her tools across the factory floor if someone had borrowed one without permission or had not returned one in the appropriate time. The heavens would then open up and the wrath of the almighty would rain down on the poor unfortunate tool borrower. I learned quickly to ask nicely and work nearby if I needed a tool. I also noticed that many of the line workers would walk across the factory floor to borrow a tool from Debbie. Being a huge proponent of Lean Manufacturing I quickly noticed this waste. I went to the plant manager and senior engineer to request that each line worker be provided a set of tools or at least a set of tools to be stored at each station for that station's activity. "No, that will cost too much and they will just take them home," was the response. I then showed them some math that I've used many times to make my point.

> If one wastes just 1.5 minutes per hour, that is 60 minutes per week, or one full hour of waste per week. The wasted time chasing tools was much greater than this at this factory. The annual waste, minus two weeks of vacation and some holidays, was about 48 hours of waste per year per person. That adds up to a lot of wasted labor dollars. I have to hand it to these two guys. They quickly caught on and the three of us went up to the local hardware store that same day, right after lunch. Each worker was then issued a complete set of tools appropriate for the job and each was then responsible for their own tools if one came up missing. Their solution was to follow Debbie's example with each choosing a color for their tools, from hot pink to Irish green.

Alright, back to SPC and consistency. The consistency was measured by an experienced assembly worker testing each reel by hand to see how it felt. About every tenth reel he would spool with fishing line and go outside and cast it. This was all touchy-feely, although the guy was really good at it. Along with engineering and their mechanic we were able to purchase a machine that we could connect the reels to and it would measure the drag of the operation of the reel. I then taught the engineers how do to manual SPC calculations, so that they could process the data into something meaningful. After they had accumulated enough data, they then developed a drag specification for each type of fishing reel. How the guy thought the reel felt no longer applied. He was still the tester and put the reels into the machine and measured the drag. They were still testing every reel when I finished my contract. I could not quite get them to believe in sampling. They just put all 100% of the reels into subgroups of five for their SPC activity. Oh well.

I was successful because the CEO was extremely happy! They really did have more consistent reels and their factory was more efficient. SPC was used to develop a reel drag specification and to monitor the processes for trends of the reel drag data. My apologies to Dr. Deming and Dr. Shewhart, I just could not get sampling instituted at this factory.

Constant Battles

There are often constant battles between various disciplines of management. Quality versus production is the most well known. Then there is engineering versus quality, maintenance versus scheduling, and so on. See Dr. Deming's point number nine. These battles or turf wars are nothing new in western manufacturing cultures.

- The production manager sees their job as getting product out the door.
- Too often, the quality manager operates out of a book and cannot grasp the concept of pragmatism.
- Engineering too often lives in a theoretical world.
- Another problem is that typically in a manufacturing organization, the quality manager is outranked by production or operations. To understand why this is an issue, review the balanced scorecard methodology.
- Scheduling wants the production line to run, regardless of the situation.
- Maintenance will shut down a machine for preventive maintenance without discussing with production or scheduling to find a mutually beneficial time.

I've personally experienced these examples of barriers and lack of teamwork. Why do I bring this up here? Because I've seen each of these disciplines use the SPC control charts to support their position.

- Production looks at the charts and says, "Everything is fine."
- Quality looks at the charts and data and says, "There are defects."
- Maintenance looks at the charts and says, "It is time for maintenance."
- Engineering looks at the charts and says, "Who changed what I developed?"
- Scheduling looks at the charts and says, "I need the product."

My point here is the control charts should not be used as supporting material in turf battles. Management should be working as a team of equal players.

- Sometimes production needs to accept that maybe there is a problem.
- Sometimes quality needs to take a more pragmatic approach to issues.
- Engineering should understand that things do change.
- Maintenance should be flexible and consider doing maintenance after normal hours or on Saturday.
- Scheduling must understand that sometimes a production line or machines just have problems.

I hope I'm being equally critical of the various disciplines. I'm a quality guy. But I see so few of my peers that really understand quality. A quality professional is one that really needs to understand quality intuitively and in a pragmatic sense in order to be effective. There are some industries where quality must be as perfect as possible. Drug sterility is an example. Aerospace must have some level of perfection for some parts of the plane since it is hard to find mechanics at 35,000 feet. But consider for all the stringent levels of quality in the automotive industry as guided by TS 164949, there sure are a lot of auto and truck recalls. Is it possible that all those procedures really don't work all that well?!

How about one more war story?

Machining Metal Parts

Let's consider some metal parts, in this case bronze, that are manufactured using a computer numerical control (CNC) mill machine. A CNC mill has multiple cutting tools that are specific to the part being made. The mill will cut on multiple axes often at the same time. Complex parts are manufactured more accurately and faster than in the previous era where the metal machining was controlled by hand by an operator. There are still operators running the CNC mills. Their function is to use the correct software program, set up with the correct cutting tools, mill the correct alloy, monitor the process, and periodically measure a part. In this example, the various measurements were fed into the SPC program on a computer.

The process had been running this part periodically for a number of years so there was a lot of data in the SPC software. At the start of a particular production run of bronze parts, there was a clear and distinct change in the variance as shown on the control chart. While still in control and specification, the process was using virtually the entire six units of standard deviation range on the control chart whereas in the past it used about two and a half units of standard deviation. The operator stated this had just started a couple of weeks before.

Well, according to the supervisor the operator was to blame. He/she did not set the machine up properly or used the wrong tooling. However, that turned out not to be the cause as it was found the setup was absolutely correct. Then, the blame was put on using tools with excessive wear. But, the tools were found to be new or newly refurbished. Then, the CNC mill itself was considered at fault. But, since the machine was producing other parts with no significant changes in variance, this could not be the cause. Since the parts were measured by multiple people, including the process engineer, the measurements taken were considered to be quite accurate. Can one start to see what is happening? First, the operator was blamed (manpower). Then, the cutting tools (machine). Then, the machine itself was the culprit. Measurement was ruled out. What was left?

Well, since the process was in control and in specification, the decision was made to continue to manufacture this part and similar parts by scheduled lots. Periodically, the CNC mill was changed over to make other parts which involved different tooling and types of metals such as aluminum. However, since this part type was commonly used by the assembly process in another department, the parts continued to be manufactured on a regular basis.

During a coffee break conversation between me, the CNC milling process engineer, and the production manager of the assembly process, this situation with the increased variance came up. The assembly production manager casually stated, "What have you guys done over there? I've been scrapping a lot of parts XYZ the past couple of months." Of course I was paying close attention, while not appearing to be overly excited. At minimum, I would have a new example for my quality class. After rather thorough discussion and admission by the process engineer that variance had increased, but insistence that there was no problem since the CNC milling process was still in control and specification, we decided to go out to the machine shop and take a look. An informal brainstorming session that included the mill operator and a mechanic then followed.

My question was, as always, "What has changed? Something had to have changed?" Always having the cause and effect root cause tool in my mind, I could intuitively rule out environment, so I asked if the material had changed. "No, it's the same stuff we always use," came the response from all. Well, really? How do we know that, I wondered? As the others went back to their normal activities, I started checking the SPC data. This organization kept the SPC data on the PC associated with the CNC mill and was not accessible from a central database. Since the mill machined several different parts, there were several different control charts and data. I checked the current Cpk, which was still greater than 1.0. I then went back a few months and found the Cpk was about 1.4. I was then able to gradually move forward in time and found just like the operator reported, there was a sudden change in the process variation. It wasn't noticeable to the engineer or machine shop supervisor because they were still looking at the total process Cpk, which was calculated based upon thousands of measurements. I isolated to the recent production batches, and then specific periods of time. That is how I saw the dramatic change in the Cpk.

I then went up to the materials department. Checking with the metals buyer on this particular size of bronze rods that were used to mill the part, I got an immediate response, "Yeah, I changed the supplier about a month or so ago. I got a really good deal and saved a lot of money!" Therein was the root cause. Bronze of lower quality was bought that was sold to be of the same type and quality of alloy, but it wasn't. The information was presented to the operations manager, who required materials to buy a batch of the bronze rods from the previous supplier. Sure enough, when this bronze was used, the variance immediately decreased and was back to about a total of two and a half units of standard deviation. So what did we learn here?

- Again, the control charts don't lie! If the measurement is accurate and variance increases, then there is some cause.
- Believe the control charts and don't forget all the data that is provided, including the Cpk. Use the SPC data as a problem-solving tool when necessary.
- Trust the people doing the work. They really want to do a good job and they usually know when there is a problem, even if they don't quite understand it. This machine operator later thanked me for listening to him when he said that there was a problem somewhere.
- Remember the six parts to a process. The root cause of the problem is always in one of the six.
- Following Dr. Deming's point number four, don't buy parts or materials on price alone.
- Read Dr. Phil Crosby's *Quality Is Free* as published by McGraw-Hill to get a better understanding of the real cost of quality. In this example, there was a dramatic increase in scrap and wasted labor downstream from the machining process in the assembly department. Also, some of the labor used in the machining process was also wasted since some of the parts were scrapped. Materials did not save the organization any money!
- Also, remember Taguchi's Loss Function when doing problem solving.

In Summary

There are some issues that cross cultural boundaries. I once asked another consultant who did a lot of work in the Middle East and Indian Ocean rim, what was the biggest weakness in ISO 9000 quality systems? Without hesitation, he replied, "Training!" I too have found that same weakness. Although my auditing has been limited to North America, I seldom found an organization where I did not have a finding in the area of training. It is just a weak link around the globe. Another area of confusion that crosses cultural boundaries is preventive action. I'm glad this requirement has been removed from ISO 9001:2015. I wrote more major findings on preventive action, or actually the lack thereof, than any other requirement. This covers audits in general industry, aerospace, medical device, and automotive industries. Very few people really understand preventive action which includes registrar auditors and regulatory agency auditors. Yet with all this confusion about what is meant by preventive action, it is still a requirement in many standards.

Remember that there are differences as one crosses cultural boundaries. Then, people in your own culture are different. What is common for one may be difficult for another. I remember working with a manager that clearly thought he was the model everyone should follow. When the Meyers-Briggs personality profile was given to the entire management staff of our organization, he stated, "I guess everyone should be ISTJ like me." When the room roared with laughter, he was clearly confused. He just didn't get it. (By the way, I am an INTP.) Consider bringing a New Yorker, a Californian, a Mississippian, and someone from Nebraska together to do a project. I guarantee there will be cultural chaos. The New Yorker will be in a hurry and pushy. The Californian will show up at 11 a.m. The Mississippian will yawn and stretch and wonder what the hurry is. The Nebraskan will be wondering what is wrong with the others. What I'm trying to point out here is that people are different and people from different cultures are different. Don't expect everyone to be like you. Don't put your work ethic on others. Why is this in an SPC book? Well, because as I have shown in the last war story, SPC can be used in a problem-solving situation. Remember the manpower portion of the cause and effect diagram. Sometimes, when the cause is people, it is not really their fault. They just do things differently.

6

Now What?

Opportunity is missed by most people because it is dressed in overalls and looks like work.

Thomas A. Edison

Well, at this point, you have people trained in SPC. Whether at your organization, at another organization, or just training for the reader, it must be used. Training is quickly forgotten if not used soon after the training is provided. Use it! For those that used the training methods in conjunction with the installation of an SPC system, start entering your data. It takes a lot of data to start to get a good picture of a process. Those that read this book and those that are trained by the methods in this book should get access to SPC software. Even if one has to start out using fictitious data, or data that is not part of a real process, put it into the software and start processing the data. Another alternative is to purchase an SPC calculator. This can be used to facilitate the creation of manual control charts and can be used to calculate the Cp or Cpk values. Then, since the calculations are demonstrated in this book, one can also do the calculations with a regular calculator. But, the most important thing to do is to start using the knowledge. With a new SPC system you will start to establish a baseline as you enter data. Monitor the process or processes and watch the control chart pictures evolve.

Let's consider the people that have been trained whether at your organization or at another. Start to observe the control charts. Understand these charts are a picture of the process. These pictures tell you a lot of information at a glance. Remember in Chapter 1, I stated that when I arrived in the morning and I turned on my PC to view the second and third shift control charts, I could accurately predict their approximate case output and level of quality. I isolated the SPC data to those shifts and had a picture. A glance at that picture told me all I needed to know.

Once there is sufficient data, use the control charts, as well as the Cp and Cpk values for improvement projects. Start looking at the control charts and see the common cause variation. Seek potential improvement projects. Develop action plans. Initially one will work on the "low-hanging fruit." These are the easier projects. These will provide the greatest value for your use of resources. As one goes on to further reduce common cause variation or choose continuous improvement processes, remember the law of diminishing returns. This means there will usually be less value gained, for an equal amount of resources used as when the easier projects were dealt with. Conversely, to get as much value as was obtained in the easier projects or an equal reduction in variation, there will need to be a great expenditure of resources. For problem project management and to measure improvements, understand how to go back and isolate a period of time and observe the picture of the control chart in that period of time to compare with the current state. Remember, the problem or continuous improvement primary root cause will be in one of the Cause and Effect categories: machine, method, material, manpower (people), measurement, or environment. As one goes through the problem solving or project management activity, look at the pictures depicted on the more recent control charts and look for improvement. Develop an intuition to understand what has happened in seconds. See the variation. See the location of the subgroups. See the Cp and Cpk values. As a team proceeds through a continuous improvement activity, remember SPC and other associated tools to use with Six Sigma project management or other project management methods such as 8D, Brainstorming, 5-Why, Root Cause Removal, Cause and Effect, and other quality tools. Be careful of secondary causes that aren't the actual primary root cause. Some examples include:

- Too often people are blamed when it is not their fault.
- Don't do a 3-Why analysis (there really is no such thing) rather than a 5-Why analysis because it is easier.
- Always do a thorough Cause and Effect Diagram completing a thorough Fishbone, with many branches of bones from other bones, when using this too.
- Understand that the true or primary root cause is usually difficult to determine.

When observing control charts, always look for assignable, special cause variation. Most importantly, see if there is any problem variation as I've described and defined in this book. Understand when there is a problem and when there is not. Remember to think of the control chart as a picture of the process. Take action on the problem variation as soon as possible. Stop the process if needed, such as when a control chart looks like the one in Figure 4.6.

In addition, I've provided some other tools to evaluate a process for the level of conformity to specifications. I've also provided some information on general quality assurance as well as some other quality related activities. These can be used as part of your problem solving or continuous improvement activities. I highly recommend following up with reading the various referenced books that I've presented. Knowledge is everything!

If you or someone you have instructed did not do well on the problems in Chapter 7, ask yourself why. What did they not grasp? Was it something in the mode or manner of the presentation? Perhaps you as the instructor do not really understand the material? Was there not enough time provided to properly provide the material in this book? I could write many pages of problems and solutions, but I have chosen not to do that. For those that struggle, I suggest changing the numbers in the problems and trying again. Check yourself or those that you are instructing against the methodology shown in the solutions portion of Chapter 7. Most of all, keep trying.

I refer to myself as a pragmatic fanatic about product quality. That is, I believe in excellence with a common sense approach. There are some industries where perfection is not really important and there are some where it is. There are industries that accept some amounts of less than perfect, otherwise known as product variance. Other industries require a high level of excellence. Back in Chapter 3 in Part 1, we defined the word quality. All three definitions involved the customer. Truly the customer has the last word on your product's quality. They have the last word on the amount of allowable variance. They have the last word on the level of acceptable defects.

For this chapter, I'll leave you with a couple of suggestions. Become an expert in excellence. In most employment situations, the quality person is a second class citizen as compared to other disciplines. Don't let that hold you back. If you are an engineer, an operations manager, or any other discipline also become an expert. Use this book to become an expert in SPC. Read the references I've provided as well as the books by the guru's I've quoted or referenced. When you retire and look back you can say, "I was a success!" and not, "I did a decent job for 40 plus years."

7

Problems with Solutions

There is nothing so deceptive as an obvious fact.

Sherlock Holmes

Whistle while you work.

Seven Dwarfs

Problem #1

Consider the following raw data with a subgroup of five.

- Xbar = 241.5, 246.1, 239.8, 240.1, and 245.4
- R = 11, 3, 15, 12, and 4
- Rbar = 10.1
- Xdbar = 242.3
- LSL = 239.0USL = 248.0
- A_2 = 0.577

Calculate:

UCL = ?   LCL = ?   Cp = ?   Cpk = ?

Construct an Xbar control chart and plot your data.
Are there any out-of-specification products being produced?
Are there any out-of-control conditions?
Should any action be taken?

Problem #2

Consider the following data:

- UCL = 11.34   LCL = 8.51
- Xdbar = 9.925
- R = 0.87

Calculate sigma.

Problem #3

Consider the following data:

- USL = 0.95   LSL = 0.65
- Xdbar = 0.72
- Sigma = 0.03

Calculate the UCL and LCL.

Problem #4

Consider the following data:

- USL = 42   LSL = 35
- Xdbar = 39.76
- Standard Deviation = 0.9

Calculate Cp and Cpk.

Problem #5

Consider the following data:

- USL = 3420   LSL = 3180
- UCL = 3410.1   LCL = 3195.3
- Rbar = 157.6

Is this process producing all products within specification? Prove your answer.

Problem #6

With a monthly defect rate of 5.742 defects per 10,000, what is the monthly defect rate in parts per million (ppm)? Is this defect rate good or bad?

Problem #7

Consider the following data:

- The subgroup averages are 72, 81, 73, 74, 72, 74, and 79.
- The subgroup ranges are 3.4. 3.7, 4.1, 2.2, 3.6, 3.5, and 2.9.
- The subgroup size is 5.
- $A_2 = 0.577$.

Calculate Xdbar, Rbar, the UCL, and the LCL.

Problem #8

Consider the following Xbar control chart for a shift of production.

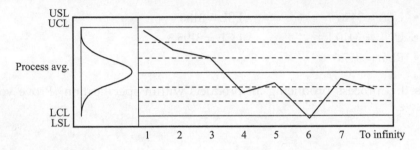

What do you see?

    A. An out-of-specification subgroup
    B. An out-of-control subgroup
    C. Problem variation
    D. Excessive variance
    E. A trend to out of control
    F. Nothing of great concern

Problem #9

Considering your answer for #8, what action would you take?

Problem #10

Consider the following Xbar control chart for a shift of production.

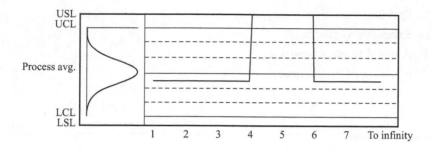

Explain what you think has happened to produce this control chart.

Problem #11

Considering your answer for #10, what action would you take?

Problem #12

Consider the following Xbar control chart.
 What do you see?

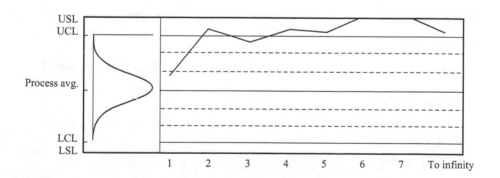

A. An out-of-specification subgroup
B. An out-of-control subgroup

C. Problem variation

D. Excessive variance

E. A trend to out of control

F. Nothing of great concern

Problem #13

Considering your answer for #12, what action would you take?

Problem #14

Which of the following is a measure of the greatest variance in a process?

A. UCL = 1.2, LCL = 0.6

B. Standard deviation = 0.057

C. R = 0.8

D. σ = 0.09

Problem #15

Which of the following is a measure of the greatest variance in a process?

A. Sigma = 10.2

B. Standard deviation = 11.9

C. R = 15

D. UCL = 15.9

Problem #16

Consider the following information: LSL = 102, UCL = 129, USL = 130, LCL = 118. What do you conclude, and what action would you take?

Solution #1

It should be clear that Xdbar does not need to be calculated. It is given. Remember Xdbar represents the entire process average. Taking the given subgroup averages and calculating a process average would be incorrect for this problem.

In addition, the given data for R is not needed. It is not relevant for this problem.

UCL = ? This is a multistep problem. Based on the information provided, the formula
UCL = Xdbar + (A$_2$Rbar) will be used.

UCL = Xdbar + (A$_2$Rbar)
    = 242.3 + (0.577 × 10.1)
    = 242.3 + 5.83
UCL = 248.1

LCL = ? Following a similar process as calculating the UCL and using the formula
LCL = Xdbar − (A$_2$Rbar)
    = 242.3 − 5.83
LCL = 236.5

Cp is the specification or tolerance width divided by the process width.
Cp = (248.0 − 239.0)/(248.1 − 236.5)
    = 9/11.6
Cp = 0.78, which is less than 1.0

C$_{pk}$ = min [(Xdbar − LSL) / 3σ, or (USL − Xdbar)/3σ]
First, calculate sigma, which is UCL − LCL divided by 6 or (248.1 − 236.5)/6 or 1.93.
Then, 3σ = 1.93 × 3
    3σ = 5.8

Filling in the values in the formula:

Cpk = min [(242.3 − 239.0)/5.8 or (248.0 − 242.3)/5.8
    = min 0.57 or 0.98
Cpk = 0.57, which is well under 1.0

Now, construct an Xbar control chart. Note this is the first time we have used the subgroup averages that were given.

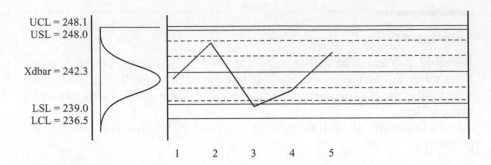

UCL = 248.1
USL = 248.0

Xdbar = 242.3

LSL = 239.0
LCL = 236.5

Based on the Cpk of 0.57, there is definitely a significant amount of out-of-specification products being produced on this process. For this shift, look at subgroup #3, which is just slightly out of specification. So there are out-of-specification products produced on this shift and on this entire process.

However, this shift is in process control.

There is problem variation since the process is significantly wider than the specification. Do the specification limits reflect the actual need or can these be changed to reflect the process? If so, engineering should evaluate new specifications that are more realistic for the process. Are the specification limits required by the customer? If so, then there is a need to reduce the total variation or common cause variation which is action on the process as a whole. This will be a significant and potentially costly process improvement.

Solution #2

Again, as in problem #1, there is data that is not needed. One should be able to recognize this. All that is needed are the control limits.

(UCL − LCL)/6 = sigma
(11.34 − 8.51)/6 = 0.47
Sigma or standard deviation from the process mean or σ = 0.47

Solution #3

The specification limits are not needed for this calculation. Again, the student must recognize what data is relevant.

UCL = Xdbar + 3σ   LCL = Xdbar − 3σ
UCL = 0.72 + (3 × 0.03)   LCL = 0.72 − (3 × 0.03)
UCL = 0.72 + 0.09   LCL = 0.72 − 0.09
UCL = 0.81   LCL = 0.63

Solution #4

Cp = Specification width/process width
Cp = (USL − LSL)/(UCL − LCL)
UCL = Xdbar + 3σ UCL = 39.76 + (3 × 0.9)   UCL = 39.76 + 2.7 UCL = 42.46
LCL = Xdbar − 3σLCL = 39.76 − (3 × 0.9)   LCL = 39.76 − 2.7LCL = 37.06
UCL − LCL = 42.46 − 37.06 = 5.4
USL − LSL = 42 − 35 = 7
Cp = 7/5.4 = 1.3, which is greater than 1.0

As in problem #1, and using data from the above Cp calculation,

C_{pk} = min [(Xdbar − LSL)/3σ, or (USL − Xdbar)/3σ]
Cpk = min [(39.76 − 35)/3 × 0.9, or (42 − 39.76)/3 × 0.9
Cpk = min (1.76), or (0.83)
Cpk = 0.83

The process is producing product that is out of specification on the upper side.

Solution #5

The simplest and fastest way to get a general idea is to use the goal post methodology. Doing a quick sketch of the data as follows:
The process is producing product within specification and is roughly centered.

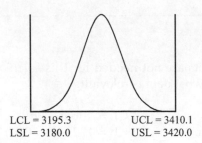

LCL = 3195.3   UCL = 3410.1
LSL = 3180.0   USL = 3420.0

Then one could calculate Cp and Cpk, which is more than is needed for this problem.

Cp = (USL − LSL)/(UCL − LCL)
Cp = (3420.0 − 3180.0)/(3410.1 − 3195.3)
Cp = 240/214.8
Cp = 1.12

This value is greater than 1.0, so the process has the potential to meet specification. Not overly great, and there is little margin for error.

However, to exactly answer the question, Cpk must now be calculated.

C_{pk} = min [(Xdbar − LSL) / 3σ, or (USL − Xdbar)/3σ]
Xdbar is the midpoint of the process.
To calculate first find this value: (UCL − LCL)/2 = 214.8/2 = 107.4 = 3σ range
Then, UCL − 107.4 or LCL + 107.4
3410.1 − 107.4 = 3302.7 or 3195.3 + 107.4 = 3302.7
This is the midpoint of the process or the process average or Xdbar = 3302.7.
To calculate the standard deviation from the process average:
(UCL − LCL)/6 or (3410.1 − 3195.3)/6 = 35.8

The standard deviation or sigma or σ = 35.8 or use the 3σ range from above or 107.4.

Back to our formula; C_{pk} = min [(Xdbar − LSL)/3σ, or (USL − Xdbar)/3σ]
Cpk = min [(3302.7 − 3180)/107.4, or (3420 − 3302.7)/107.4]
Cpk = min 122.7/107.4, or 127.3/107.4
Cpk = min1.14, or 1.09
Cpk = 1.09

The process is truly close to being centered in the specification limits as shown by the choice of the minimum of 1.14 or 1.09. All products are within specification, although there is not a significant amount of room for error.

Solution #6

This problem is not an SPC problem, but is added since ppm is so often referenced in the quality world.

5.742 per 10,000 or 10^4 is how many in ppm or parts per 10^6? Just multiply by 10^2 or 100 and get 574.2 ppm. Is this good or bad? It depends on what industry is involved and what product, and most of all the customer expectations.

Solution #7

Since one is calculating the process average and process range average, then one has to assume just a small portion of production is being examined. The value for A_2 is given in problem #1.

Xdbar = (72 + 81 + 73 + 74 + 72 + 74 + 79)/7   Divide by seven since there are seven pieces of data.

Xdbar = 75 for this period in time of the process.

Rbar = (3.4 + 3.7 + 4.1 + 2.2 + 3.6 + 3.5 + 2.9)/7
Rbar = 3.34 for this period in time of the process.

UCL = Xdbar + (A$_2$Rbar)   LCL = Xdbar − (A$_2$Rbar)
UCL = 75 + (0.577 × 3.34)   LCL = 75 − (0.577 × 3.34)
UCL = 75 + 1.93   LCL = 75 − 1.93
UCL = 76.93   LCL = 73.07

Solution #8 and #9

A. There are no out-of-specification subgroups.

B. Subgroup #6 is slightly out of control.

C. I do not see any problem variation.

D. The variation does range from one control limit to the other. This can be considered excessive. However, I would still not take action on the process as I do not see what I consider to be problem variation. However, I would get more data by measuring more samples.

The Cpk value may be low. Again, I would look for causes by getting more data before taking action.

E. The process did trend to the lower control limit. However, it then came back toward the center at subgroups #7 and #8 is only one standard deviation from the process average. As in answer D, I would get more data and would take no action.

F. Since I do not see any problem variation, I would not have a great concern. But, as in answers D and E, I would get more data.

The best answer is F, although D is acceptable as long as one understands it is necessary to get more samples to measure in order to have more data to evaluate, before taking any action.

Solution #10

This is a chart with a data entry error. Subgroup #5 is well off the chart and the subgroup is off by at least two orders of magnitude. It is so far beyond the upper control limit and the process average that this subgroup value cannot be valid. Notice that this subgroup value has caused the rest of the values to flat line.

Solution #11

All that has to happen is for someone with administrative access to the software to edit or delete the value for subgroup #5. If this was manual data entry, then the numerals are likely correct, but the decimal point is missing or entered in two or three places incorrectly. Just move the decimal point to the correct location. If this is automatic data entry, then it is machine error or improper use by the operator and the value should be deleted.

Solution #12 and #13

A. Yes, there are out-of-specification subgroups at the upper level at subgroup #6 and #7.

B. Subgroup #2, #4, #5, #6, #7, and #8 are out of control on the upper control side.

C. I see some problem variation and I'm concerned, but I would not immediately start changing the process. Depending on what was being produced, I might or might not stop the process. I would get additional data by measuring more samples.

D. There is no excessive variance. Total variance on this chart is about four units of standard deviation.

E. There is no trend. The process is just out of control on the upper side at this point.

F. Yes, there is concern as stated in answer C.

The best answer is C.

Solution #14

A. The standard deviation can be quickly calculated as 0.1. This is the correct answer.

B. This standard deviation is less than answer A.

C. Not a relevant answer. The range of one subgroup is not a measure of process variance.

D. This standard deviation is less than answer A.

Solution #15

1. Less standard deviation than answer B.

2. This is the correct answer.

3. As in problem #14, this is not a relevant answer.

4. This is not a measure of variance.

Solution #16

To graphically view the data, construct the goal posts.

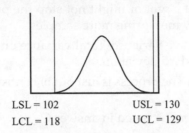

LSL = 102   USL = 130
LCL = 118   UCL = 129

One could also quickly sketch a control chart, which is essentially show-ing these same values horizontally. I just have a preference for the goal post methodology.

The conclusion is that the process is within specification. Cp could be cal-culated, but it will be well above one point zero as can be visualized by the goal post sketch and estimating the specification spread over the process spread. Because the process is shifted to the upper end of the process with the UCL just below the USL, the Cpk value will be just slightly above one point zero demonstrating, as shown by the goal posts, there is little margin for error before going out of specification.

Once again, I would be hesitant to take any action on the process since there is not any evidence of problem variation. I might evaluate why the pro-cess is running at the high end of the specification, but again, I would take no immediate action.

8

If It Ain't Broke ... Don't Fix It

I like the dreams of the future better than the history of the past.

Thomas Jefferson

Learning is experience. Everything else is just information.

Albert Einstein

What I hope most is that the reader or students involved in the use of the methodologies provided in this book have a foundation in which to use SPC. Even if one just understands the basics of SPC and can recognize a sudden assignable or special cause variation event, then that is success. However, more than that, I hope the professionals that read this book or are trained in the information provided are able to understand the concept of problem variation. I hope that the information presented in Chapter 6 will be used to move toward excellence in one's career and activities.

Then remember as I have repeated over and over: When changes are made to a process, things will get worse before improvement is seen! Only make changes when absolutely necessary! Learn to recognize problem variation. Only take action when there is problem variation, unless you are working as part of an improvement project. Understand that an improvement project will probably not show immediate success as shown on the control chart or by the Cpk. Then train your operators not to make unauthorized changes to the process. If things go crazy, then have them contact the appropriate professional be it management or engineering. Train and understand how recognize problem variation. Therein, possibly hiding somewhere, will be your problem.

Quality is never an accident. It is always the result of intelligent effort.

John Ruskin

When written in Chinese, the word "crisis" is composed of two characters—one represents danger, and the other represents opportunity.

John F. Kennedy

The best leaders ... almost without exception and at every level, are master users of stories and symbols.

Tom Peters

Excellence is possible if you:
Care more than others think wise.
Risk more than others think safe.
Dream more than others think practical.
Expect more than others think possible.

Glossary

With a little help from my friends.

John Lennon and Paul McCartney

In statistics and statistical process control, the nomenclature is critically important and often confusing. In addition, not all these terms are used in this book, but rather are considered important for the reader to be familiar with.

action: in the manufacturing or service world, action is taken to correct a minor, usually one-time, nonsystemic event. *See* corrective action which is different.

alpha or type I error: when a good lot or batch is incorrectly rejected. This is the producer's risk. *See* beta error.

assignable cause variation: this type of variation will have a specific root cause. It is not a natural part of the process and is readily seen on a control chart. It is usually unpredictable. One will usually want to eliminate assignable cause variation. This is because it is an anomaly and not really part of the process. The root cause of the assignable cause variation will be found somewhere on the Ishikawa or fishbone diagram.[1]

attribute or attribute defect: a quality or characteristic associated with a specified requirement. It is an expression of the number of articles counted, rather than a measurement. Is the characteristic present, yes or no?

bar graph: *see* histogram.

batch: *see* lot.

beta or type II error: the mistake of accepting a bad lot or batch, usually due to improper sampling. It is known as the buyer's risk. *See* alpha error.

capability: a measure of the extent to which the product meets its specified operational requirements during use, given its dependability during the period of use.

cause and effect diagram: also called a fishbone diagram or Ishikawa diagram. It is a root cause analysis tool used to break down potential causes and potential sub-causes into categories of people, methods, machines, materials, measurements, and environment.

common cause variation: think of this as the background "noise" in a process that is seen on a control chart. It is the normal variation in a process. The world is not perfect. Common cause variation always exists and is difficult to reduce. It can never be eliminated. There are many causes in the process that contribute to the common cause variation.

component: any material, piece, part, or assembly used during manufacturing that is intended to be included in the finished product. This does not include "manufacturing materials."

conformance: an affirmative indication or judgment that a product or service has met the requirements of the relevant specifications, contract, or other requirement; also, the state of meeting the requirements.

conformity: the fulfilling by an item or service of specification requirements.

control chart: a graphic display of the state of control of a process. It is often used to monitor whether a process is within the established specification limits, while in a state of process control. Control charts help one to concentrate on the *common causes* for process variations and help to eliminate the tendency to focus on *special causes* for process variation. Control charts can also assist in determining the capability of a process. Control charts are a form of SPC. Xbar, R Charts, and *p* Charts are examples.

corrective action: is the systemic action taken to prevent or minimize recurrence of a discrepancy. The act of permanently removing a circumstance that has caused or may cause a deficiency in the product, project, or service. *See* action.

defect: an individual failure to meet a single requirement.

defective: a unit of product which contains one or more defects.

Deming cycle or Deming wheel: originally the Shewhart cycle. It is composed of four stages: plan, do, study (check), and act. It is a never-ending cycle, focusing on continuous improvement.

efficiency: the ratio of useful output (work, plus losses or wasted energy) to input energy.

fishbone diagram: *see* cause and effect diagram.

frequency: the number of complete cycles that take place in unit time.

grand average: *see* process average.

histogram: captures visually the variation in a process and generally displays a pattern. It is a graphical representation of the variation in a set of data. It shows the frequency or number of observations of a particular value or within a specified group. Also known as a bar graph.

inspection: includes activities, such as measuring, examining, testing, gauging one or more characteristics of a product or service, and comparing the resulting data with specified requirements to determine conformity.

lot: a specific group of products, components, or materials that has uniform character and is produced according to a single manufacturing order, during the same cycle of manufacture.

manufacturing material: items used to facilitate the manufacturing process, but which are not included in the final product. Some examples are cleaning agents, lubricants, and spacers.

normal bell curve: a bell-shaped curve representing amounts of data at any specific point on the curve. Also known as the normal distribution. It is a basic component of SPC. The mean, median, and mode are all at the same point on the curve.

***p* chart:** the "*p*" refers to percentage or proportion of defects found in the inspected sample. It is an "attribute" control chart that displays the amount of occurrences of a characteristic over time. It does not require any actual measurements.

Pareto* diagram: a powerful tool used to analyze attributes data collected in check sheets. It is a form of "histogram" where the vital few are separated from the trivial many. Three scales are used; characteristic at the bottom, frequency on the left side, and percentage on the right side. Characteristics are grouped from the largest frequency to the smallest, showing the relative magnitude of each. A cumulative frequency is then drawn, which is used to determine which characteristics are 80% of the problem. While not discussed in this book, it is an interesting phenomenon to study.

Pareto principle: most effects come from relatively few causes. In quantitative terms, 80% of the problems come from 20% of the causes; equipment, components, operators, and so on. The 20% that causes the most problems is known as the *vital few*, while the remaining 80% is known as the *trivial many*. Vilfredo Pareto was an Italian economist, who in 1906 observed that 80% of the land in Italy was owned by 20% of the people. His legacy is considered very profound by many. His 80/20 concepts in economics have been applied in many other fields.

parts per million (PPM): refers to defects or errors per million which makes a defect rate of different processes more easily comparable.

problem variation: as defined in this book, this is the only type of variation where action or corrective action must be taken on the process. This is contrary to other SPC writings as discussed in this book.

process: a sequence of events or activities completed by a person, group, or equipment. It is normally a combination of people, methods, machines, materials, or environment.

process average or grand average or Xdbar or X double bar: the average of a process as measured over time.

process capability: the ability of a process to produce an output that meets defined specifications.

process capability index: "Cp" and "Cpk" are used when the process is in statistical control and when the process forms a normal distribution curve. Cp is calculated to measure potential capability. It assumes the process is centered between the upper and lower tolerance limits.

---

* Vilfredo Pareto (1848–1923): An Italian economist, engineer, sociologist, political scientist, and philosopher who developed the Pareto principle or the 80/20 Rule. He based his principal on studies of wealth or income distributions.

$$Cp = \text{Specification Tolerance(Upper Limit-Lower Limit)} / 6\sigma$$

If Cp equals one, the process is barely able to meet the specification. A Cp value of greater than one represents a process that is able to meet the specification, with the higher the Cp value, the better. A Cp of less than one represents a process not capable of meeting the specification. A Cp value of 1.67 is considered excellent.

Cpk takes into account the lack of centering of the process between the upper and lower tolerance or specification limits. This is a more useful measurement since processes rarely remain fixed at the center.

$$\text{Cpk=the lesser of}$$

$$[\text{upper specification limit-X double bar}]/3s$$

or

$$[\text{X double bar-lower specification limit}]/3s$$

A Cpk of greater than one shows capability of meeting specification. A Cpk of less than one shows a process not capable of meeting specification. A Cpk of 1.33 is considered very good.

quality: the perception of what a customer believes he/she wants and expects. Or the totality of features and characteristics of the product or service that reflect on the ability to satisfy the stated or implied need of the customer.

quality assurance: all of the planned or systematic actions including development, processing, inspecting, testing, distribution, sales, and service needed to provide confidence that a product or service will satisfy the given requirements. It is a total systems approach and must be an integrated effort by all departments.

quality circles: a cross-functional team composed of members both directly or not directly involved with a process that meet regularly to improve quality and efficiency.

quality control: is a basic quality program that is part of an overall quality assurance system. Quality control typically evaluates raw materials, components, and finished products for defects. Some systems use quality control to do in-process evaluations.

quality engineering: that branch of engineering which deals with the principles and practice, of products and services, through the principles quality assurance and control.

quality system: the organizational structure, responsibilities, procedures, processes, and resources for implementing quality management.

range average: the average of a group of range values in a process. Values are plotted on the range chart.

range or R: the difference between the high and low measurements in a subgroup. These values are then plotted on the range chart.

range or R chart: a type of control chart used to evaluate variable data of subgroup ranges.

relative quality: the relative degree of excellence of a product or service.

reliability: the probability that an item will perform a required function under stated conditions for a stated period of time.

root cause: the fundamental, underlying reason for a problem. Focusing on a root cause of a problem generally leads to a permanent fix. It is the ultimate reason for nonconformity in a process.

run chart: a graphic display that shows a measurement against time, with a reference line to show the average of the data. It is similar to a control chart, but does not show process control limits and may not show tolerance or specification limits.

sigma: a statistical measure of variation around the mean of a distribution or a *standard deviation*. It is the amount that a controlled process can be expected to vary from its average or mean performance. Referred to as sigma units or units of standard deviation from the process mean or average. It is represented by the symbol σ.

Six Sigma: a Six Sigma process is one that produces 3.4 defects per million opportunities. While this level is rarely achieved, it has developed into a project management methodology for reducing defects and product variation.

Six Sigma range in statistical process control: represents when 99.73% of the output is within +/− three sigma or sigma units (3σ) from the centerline or process average of the control chart. Not to be confused with Six Sigma methodology, whose practitioners earn various colors of belts: black belts, green belts, yellow belts, and so on.

special cause variation: *see* assignable cause variation.

specification: a formal description of the desired state of a condition or process. It is sometimes considered as the voice of the engineer. Typically, the condition is written. It may be developed internally or be received from the customer.

specification range or tolerance: consider this to be the theoretical range that the process is designed by engineering to live. Specifically, it is an allowable deviation from a standard or target that represents perfection. Once a process is introduced to the real world, specifications are often changed to meet reality.

standard deviation: *see* sigma.

statistical process control (SPC): the application of statistical techniques to the control of processes. Generally, the accumulated data is shown on a control chart or other type of graphic illustration. Sometimes, it is incorrectly referred to as *statistical quality control*. This methodology is used to "control" the process.

statistical quality control (SQC): it is the application of statistics to control quality. It includes but is not limited to SPC, sampling plans, and Pareto analysis.

subgroup average: the arithmetic mean of a group of data sampled in the same relative time period or condition. The subgroup average is then plotted on a control chart.

Taguchi Loss Function: a concept defined by Genichi Taguchi that states "any variation from the nominal or desired value creates a loss and that the loss increases with the degree of variation." A part or product can be within a specified tolerance, but the closer to the outside range of the tolerance, the greater the problems encountered.

target: *see* specification.

tolerance: the amount by which the measure of a part or component can be allowed to vary from the intended value.

variable: a measured characteristic that takes any value. Any characteristic that can be measured. Compare to attribute.

variation: the unavoidable differences among the individual outputs of a process. There are two major types; common cause and assignable cause. However, in this book, I have defined problem variation.

Xbar chart: a control chart displaying the arithmetic mean of groups of sample data over time.

X bar or Xbar: *see* subgroup average.

X double bar or Xdbar: *see* process average.

Index